**Introduction to Modern Analysis of
Electric Machines and Drives**

Introduction to Modern Analysis of Electric Machines and Drives

Paul C. Krause

Thomas C. Krause

IEEE Press Series on Power and Energy Systems
Ganesh Kumar Venayagamoorthy, Series Editor

IEEE PRESS

WILEY

Published by John Wiley & Sons, Inc., Hoboken, New Jersey.
Published simultaneously in Canada.

For general information on our other products and services or for technical support, please contact our Customer Care Department within the United States at (800) 762-2974, outside the United States at (317) 572-3993 or fax (317) 572-4002.

Wiley also publishes its books in a variety of electronic formats. Some content that appears in print may not be available in electronic formats. For more information about Wiley products, visit our web site at www.wiley.com.

Library of Congress Cataloging-in-Publication Data

Names: Krause, Paul C., author. | Krause, Thomas C., author.
Title: Introduction to modern analysis of electric machines and drives /
    Paul C. Krause, Thomas C. Krause.
Description: Hoboken : Wiley, [2023] | Series: IEEE Press series on power
    and energy systems | Includes bibliographical references and index.
Identifiers: LCCN 2022032455 (print) | LCCN 2022032456 (ebook) | ISBN
    9781119908159 (hardback) | ISBN 9781119908166 (adobe pdf) | ISBN
    9781119908173 (epub)
Subjects: LCSH: Electric machinery. | Electric driving.
Classification: LCC TK2000 .K68 2023 (print) | LCC TK2000 (ebook) | DDC
    621.31/042–dc23/eng/20220824
LC record available at https://lccn.loc.gov/2022032455
LC ebook record available at https://lccn.loc.gov/2022032456

Set in 9.5/12.5pt STIXTwoText by Straive, Pondicherry, India

Contents

Author Biography

Paul C. Krause retired from Purdue University, West Lafayette, IN, USA, in 2009 after 39 years as professor of Electrical and Computer Engineering where he won every major teaching award. He also taught at the University of Wisconsin, Madison, WI, USA, and the University of Kansas, Lawrence, KS, USA. He formed PC Krause and Associates (PCKA) in 1983 and now serves as a member of the Board of Directors. He is a Fellow of IEEE and received the 2010 Nikola Tesla Award. He has written over 100 transaction papers and five books on machines and drives.

Thomas C. Krause received the BS degree in electrical engineering from Purdue University, West Lafayette, IN, USA, in 2019, and the MS degree in electrical engineering and computer science from the Massachusetts Institute of Technology, Cambridge, MA, USA, in 2021. He is currently pursuing the PhD degree with the Massachusetts Institute of Technology.

Foreword

In their preface, the authors connect the arbitrary reference frame analysis of ac machines to Tesla's rotating field. This transformation of Tesla's rotating magnetic field to any reference frame lends itself to selecting a frame that aids analysis and provides insight for advanced control algorithm development. This flexibility is most beneficial with the rise of adjustable speed ac drives. This development has changed the relative importance of reference frame theory (RFT), and the authors make a sound case for introducing RFT early in a student's undergraduate program. The authors state and I concur from experience that the arbitrary reference frame can be applied to synchronous machines of all types and induction machines.

One of the earliest uses I made of proper reference frame selection in the analysis of ac machines occurred in the late 1970s during the energy crisis. It is then an emphasis was placed on machine energy efficiency. One approach proposed was the development of permanent magnet single-phase motors with the intent to replace single-phase induction motors. The presence of an unsymmetrical rotor dictated a rotor reference frame. By applying harmonic balance to the resulting model provided an analysis strategy resulting in accurate predictions of the total torque on the machine with its inductive accelerating component and its braking permanent magnet component.

In the late 1980s and early 1990s change in the industrial sector was beginning at great speed. With the invention of sizable power transistors and then insulated gate bipolar transistors (IGBT) dc drives, the backbone of industrial power, were challenged by low voltage source inverters. Although still in its infancy, motor control chips were under investigation with the promise of control algorithms only written about in theoretical papers. Early on, high-performance current regulation was identified as necessary to achieve the performance beyond volts per hertz operation. Initial controllers attempted to achieve high performance necessary for difficult applications included three independent proportional/integral (PI) controllers. These controllers were implemented in the stationary reference frame to negate the transformation to the synchronous frame resulting in burdening the microcontrollers of the day. But there were problems of performance and stability. Our team quickly realized the error when employing three independent PI controllers in a three-phase motor controller and developed a stationary equivalent to a synchronous reference frame PI controller.

Simultaneous with the development of high-performance current regulation was the development of field-oriented control (FOC). FOC provides dynamic control of ac machines beyond volts per hertz and comparable to dc machines. This discovery opened the door for ac drives to attack high-performance markets like spindle drives and servo drives formerly dominated by dc drives. The theory behind this control is easily described by proper reference frame selection. For induction machines aligning the synchronous frame such that the rotor flux only exists in one of the two d–q axes sets up a system with decoupled rotor dynamics – not unlike the performance of dc machines. Our team further observed selecting the synchronous frame without adjusting the reference frame angle to nullify one rotor axis' flux, but examining the constraint embedded within the rotor equations provides a model that was conducive to online adaption for changes in rotor resistance and magnetic saturation. This led to model reference and observer-based controllers still in use 30 years later.

Another advance was made possible by applying RFT online parameter identification. In our development of advanced high-performance FOC controllers heating of the stator and rotor raising their respective resistances leads to performance degradation especially at low speeds. By rotating the synchronous frame by the angle necessary to align this new reference frame with the current vector provides a means for flux control and stator resistance identification.

FOC requires detailed knowledge of the load machine. General purpose drives lack the luxury of controlling a prespecified machine. As a result ac drives incorporate a commissioning procedure to identify critical machine parameters for the controller employed. Deterministic approaches are the most prevalent and are divided into transient and steady state. Through considerable testing we concluded that transient approaches were problematic chiefly because of inverter nonlinearities. Consequently, we developed a commissioning procedure that incorporated the stationary reference frame for induction motors wherein the machine was excited in a single-phase fashion with a sinewave at a sufficiently high frequency. This approach would yield two critical machine parameters necessary for high-performance FOC machines. Here again knowledge of RFT provided an avenue for solving a fundamental problem in the evolution of ac drives.

With the heightened concern over harmonic distortion, utilities have demanded from drive manufacturers improvements in drive performance as measured by the harmonic content presented to the distribution network by the drive system. This has resulted in considerable investment into developing reduced harmonic rectifiers by drive manufacturers. Among critical functions for active front ends are grid synchronization and resonance identification and rejection. One approach incorporates nonlinear adaptive tracking filters. By employing RFT it is possible to design a nonlinear bandpass tracking filter that has unique characteristics because of proper implementation of RFT.

Invention using RFT continues in areas of motor control, motor diagnostics, grid interface, drive protection, and in applications unanticipated only a few years ago.

I know of no one better able to bring RFT technology to the undergraduate and graduate students than the authors. The contributions to RFT by the authors have a long history spanning over 50 years, and numerous students have achieved technical prominence with RFT contributing to their success; technical papers in respected journals authored or coauthored by the authors are numerous. This text is in line with the previous texts "Electromechanical Motion Devices," "Analysis of Electric Machinery and Drive Systems," "Analysis of Electric Machinery," and "Electromechanical Motion Devices," all of which are widely read and distributed.

Russel J. Kerkman   Distinguished Engineering Fellow (retired)
                                                        Rockwell Automation

Russel J. Kerkman received the BSEE, MSEE, and PhD degrees from Purdue University, West Lafayette, IN, USA, all in electrical engineering. From 1976 to 1980, he was an electrical engineer with the Power Electronics Laboratory of Corporate Research and Development, General Electric Company, Schenectady, NY, USA. He is a retired Distinguished Engineering Fellow, Rockwell Automation/ Allen Bradley Company, Mequon, WI, USA.

Preface

It has been established that the transformation to the arbitrary reference frame used in the analysis of ac machines is contained in the expression of Tesla's rotating magnetic field for sinusoidally distributed windings. The voltage and flux linkage equations can be expressed in any frame of reference by simply assigning the speed of the arbitrary reference frame. The transformation is nothing more than a means of expressing the variables that portray Tesla's rotating magnetic field from a given reference frame. This establishes a meaning to the transformation and makes it much easier to understand. In addition, this allows location of the dynamic and steady-state poles in the synchronously rotating reference frame which can then be superimposed on the instantaneous and steady-state phasor diagrams. The poles provide a direct means of visualizing motor and generator action. In previous texts, Reference Frame Theory was an optional analysis technique. In this text, Reference Frame Theory is central to the analysis of ac machines.

The electric drives area has become and will continue to be an important electrical engineering discipline. Reference Frame Theory is necessary to analyze modern electric drives and it should be introduced to the student early in their undergraduate program. We can no longer just teach steady-state analysis. We must meet the challenges of the drives area and prepare the undergraduate with modern analysis tools. This book is an attempt to accomplish this goal by using Reference Frame Theory throughout. We feel this is the future approach to ac machine analysis for the undergraduate.

The arbitrary reference frame can be used for synchronous and induction machines. The synchronous machine has an unsymmetrical rotor and therefore is generally analyzed in the rotor reference frame. For purposes of analysis, however, the synchronous and induction machines differ only in the rotor configurations, the stators are the same. Therefore, the stator variables are transformed once, rather than for each machine. Once the transformation for the symmetrical stator variables has been established and the arbitrary reference frame variables set forth, only the transformation of the rotor variables of the symmetrical induction machine is needed. This transformation is very much the same as the transformation of the stator variables. This unified and compact approach prevents repeating material and makes machine analysis easier to convey to the student.

This book can be used as either the first or second course in the power and drives area as a two- or three-hour course, depending on the depth of coverage and the area program. In Chapter 1, some of the common concepts used by most authors of machine analysis are set forth. The transformation of the symmetrical two- and three-phase stator variables to the arbitrary reference frame is covered in Chapter 2. The two- and three-phase symmetrical induction machines are analyzed in Chapter 3. The three-phase permanent-magnet ac machine and the synchronous generator are treated in Chapter 4. The voltage and torque equations for the synchronous generator are established from the equivalent circuit which is established from the work in previous chapters. This approach significantly reduces the time to obtain the necessary equations.

The dc machine and dc drive are covered briefly in Chapter 5 which provides a comparison with the drives in Chapter 6 where the brushless dc and the field-oriented induction motor drives are considered. Although the power electronic switching for the ideal drive inverter is set forth in Chapter 6, courses in power electronics and controls are not required. It is assumed that the control is working perfectly, in other words, it is not how the control is designed, it is what a well-designed control system does. This chapter is followed by Chapters 7 and 8 covering single-phase induction motors and stepper motors. Symmetrical components, which can be obtained from the arbitrary reference frame transformation, are used to analyze the single-phase induction machine. Neither the analysis of the stepper motor nor the dc machine requires Reference Frame Theory.

If the interest is in drives the first six chapters, except for the synchronous generator in Chapter 4, would be covered and, if time permits, Chapter 8 on stepper motors. If the interest is in power systems, then Chapters 6 and 8 can be omitted.

Paul C. Krause
Thomas C. Krause

1

Common Analysis Tools

1.1 Introduction

The electric machine consists of a stationary member called the stator and inside this stator is a rotating member called the rotor. The stator and rotor are generally constructed from conductive wire, iron (steel), and/or permanent magnets. For alternating current (ac) machines, the main focus of this text, the rotor is different for each type of machine, but the stators are essentially the same. This chapter introduces tools to analyze the currents and magnetic fields that flow through and about the stators and rotors of electric machines.

Since the beginning of analysis of machines, several basic tools have become more or less standard. These concepts are covered briefly in this chapter. Most are used in the analysis of the machines considered in this text. This chapter starts with phasors which is a complex-number means for analyzing steady-state ac variables and ends with two- and three-phase stator arrangements. These concepts have been covered by many authors but are necessary and warrant consideration in texts on the analysis of machines.

1.2 Steady-State Phasor Calculations

We will deal with steady-state sinusoidal variables in this text and phasor analysis is very convenient for analyzing these variables. In the early 1900s, Charles Stienmetz set forth a method of analyzing the steady-state sinusoidal variables. This method has evolved over the years with different names, for example, vector analysis, sinor analysis, and now phasor analysis; however, depending on the area of application, the phasor may be slightly different. We will define it as used in the power and drives areas, which may differ somewhat from that taught in other courses.

Introduction to Modern Analysis of Electric Machines and Drives, First Edition.
Paul C. Krause and Thomas C. Krause.
© 2023 The Institute of Electrical and Electronics Engineers, Inc.
Published 2023 by John Wiley & Sons, Inc.

The phasor is established by expressing a steady-state sinusoidal variable as

$$F_s(t) = F_p \cos \theta_{ef} \tag{1.2-1}$$

where the s subscript is used here to denote sinusoidal quantities. In the following chapters, the s subscript will denote stator variables. The sinusoidal variations are expressed as cosines, capital letters are used to denote steady-state quantities, and F_p is the peak value of the sinusoidal variation. Here, F is just a placeholder for any quantity of interest. Generally, in circuit analysis, F will be V for voltage or I for current. For steady-state conditions, θ_{ef} may be written as

$$\theta_{ef}(t) = \omega_e t + \theta_{ef}(0) \tag{1.2-2}$$

where ω_e is the electrical angular velocity in rad/sec and $\theta_{ef}(0)$ is the time-zero position of the electrical variable. Substituting (1.2-2) into (1.2-1) yields

$$F_s(t) = F_p \cos\left[\omega_e t + \theta_{ef}(0)\right] \tag{1.2-3}$$

Now, Euler's formula is

$$e^{j\alpha} = \cos\alpha + j\sin\alpha \tag{1.2-4}$$

and since we are expressing the sinusoidal variation as a cosine, (1.2-3) may be written as

$$F_s(t) = \text{Re}\left\{F_p e^{j\left[\omega_e t + \theta_{ef}(0)\right]}\right\} \tag{1.2-5}$$

where Re is shorthand notation for the "real part of." Equations (1.2-3) and (1.2-5) are equivalent. Let us rewrite (1.2-5) as

$$F_s(t) = \text{Re}\left\{F_p e^{j\theta_{ef}(0)} e^{j\omega_e t}\right\} \tag{1.2-6}$$

We need to take a moment to define what is referred to as the root-mean-square (rms) of a sinusoidal variation. In particular, the rms value is defined as

$$F = \left(\frac{1}{T}\int_0^T F_s^2(t)dt\right)^{1/2} \tag{1.2-7}$$

where F is the rms value of $F_s(t)$ and T is the period of the sinusoidal variation. It is left to the reader to show that the rms value of (1.2-3) is $F_p/\sqrt{2}$. Therefore, we can express (1.2-6) as

$$F_s(t) = \text{Re}\left[\sqrt{2}F e^{j\theta_{ef}(0)} e^{j\omega_e t}\right] \tag{1.2-8}$$

By definition, the phasor representing $F_s(t)$, which is denoted with a raised tilde, is

$$\tilde{F}_s = F e^{j\theta_{ef}(0)} \tag{1.2-9}$$

which is a complex number. We see from (1.2-8) and (1.2-9) that if we consider the complex plane and rotate $\sqrt{2}\tilde{F}_s$ counterclockwise (ccw) at the angular velocity of the sinusoidal variable, the real projection is the instantaneous sinusoidal variable. We can stop the rotation and work only with the complex number. In sinusoidal steady state with a single source, the quantities of interest in a linear system will oscillate at the same frequency but with different magnitudes and relative phases. Phasor analysis keeps the amplitude and relative phase of sinusoidal quantities and eliminates the redundant information, frequency. Phasors of all like frequencies may be added by adding the real parts and imaginary parts of each phasor. We will use phasors extensively.

The reason for using the rms value as the magnitude of the phasor will be addressed later in this section. Equation (1.2-6) may now be written as

$$F_s(t) = \text{Re}\left[\sqrt{2}\tilde{F}_s e^{j\omega_e t}\right] \tag{1.2-10}$$

A shorthand notation for (1.2-9) is

$$\tilde{F}_s = F\underline{/\theta_{ef}(0)} \tag{1.2-11}$$

Equation (1.2-11) is commonly referred to as the *polar form* of the phasor. The *Cartesian* form is

$$\tilde{F}_s = F\cos\theta_{ef}(0) + jF\sin\theta_{ef}(0) \tag{1.2-12}$$

When using phasors to calculate steady-state voltages and currents, we think of the phasors as being stationary at $t = 0$; however, we know from (1.2-10) that a phasor is related to the instantaneous value of the sinusoidal quantity it represents. In other words, the real projection of the phasor $\tilde{F}_s$ rotating counterclockwise at ω_e is the instantaneous value of $F_s(t)/\sqrt{2}$. Thus, with $\theta_{ef}(0) = 0$ in (1.2-3)

$$F_s(t) = \sqrt{2}F\cos\omega_e t \tag{1.2-13}$$

the phasor representing (1.2-13) is

$$\tilde{F}_s = Fe^{j0} = F\underline{/0°} = F + j0 \tag{1.2-14}$$

For

$$F_s(t) = \sqrt{2}F\sin\omega_e t$$
$$= \sqrt{2}F\cos(\omega_e t - 90°) \tag{1.2-15}$$

the phasor is

$$\tilde{F}_s = Fe^{-j\pi/2} = F\underline{/-90°} = 0 - jF \qquad (1.2\text{-}16)$$

We will use degrees and radians interchangeably when expressing phasors. Although there are several ways to arrive at (1.2-16) from (1.2-15), it is helpful to ask yourself where must the rotating phasor be positioned at time zero so that, when it rotates counterclockwise at ω_e, its real projection is $(1/\sqrt{2})F_p \sin \omega_e t$? It follows that a phasor of amplitude F positioned at 90° represents $-\sqrt{2}F \sin \omega_e t$.

To summarize, a sinusoidal variation can be viewed as the real projection of a rotating line equal in magnitude to the positive peak value $(\sqrt{2}F)$ of the variation and rotating counterclockwise in the complex plane at the electrical angular velocity of the sinusoidal variation. Since we are in steady state and the electrical angular velocity is constant, we can stop the rotation at any time and view it as a fixed line. This fixed line is the phasor representation of the sinusoidal quantity depicted in phasor diagrams. A phasor diagram is shown in Fig. 1.A-1. Please understand that if we ran at ω_e in unison with the rotating $\sqrt{2}F$ line, it would appear as a constant to us.

In order to show the facility of the phasor in the analysis of steady-state performance of ac circuits and devices, we will consider the following circuit elements, a resistor with resistance, R, an inductor with inductance, L, and a capacitor with capacitance, C. Thus, using uppercase letters to indicate sinusoidal steady-state variables, the voltage across a resistance may be expressed in terms of the current flowing through it. That is, with I_R given as

$$I_R = \sqrt{2}I \cos[\omega_e t + \theta_{esi}(0)] \qquad (1.2\text{-}17)$$

$$\begin{aligned} V_R &= RI_R \\ &= R\sqrt{2}I \cos[\omega_e t + \theta_{esi}(0)] \end{aligned} \qquad (1.2\text{-}18)$$

In phasor form, the voltage across the resistor is in phase with the current through it as shown in Fig. 1.2-1 [$\theta_{esv}(0) = \theta_{esi}(0)$]. Thus,

$$\tilde{V}_R = R\tilde{I}_R \qquad (1.2\text{-}19)$$

For the inductor

$$V_L = L\frac{dI_L}{dt} \qquad (1.2\text{-}20)$$

where

$$I_L = \sqrt{2}I \cos[\omega_e t + \theta_{esi}(0)] \qquad (1.2\text{-}21)$$

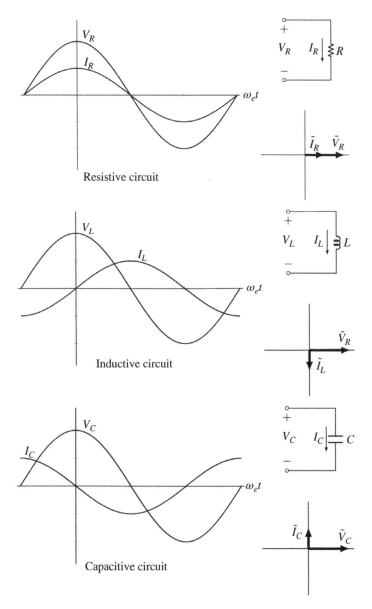

Figure 1.2-1 Waveforms of steady-state variables in resistive (R), inductive (L), and capacitive (C) circuits.

$$\frac{dI_L}{dt} = \omega_e \sqrt{2}I \cos\left[\omega_e t + \theta_{esi}(0) + \frac{1}{2}\pi\right] \tag{1.2-22}$$

Now

$$\tilde{V}_L = \omega_e L\, e^{j\frac{\pi}{2}}\, \tilde{I}_L$$
$$= j\omega_e L\, \tilde{I}_L \tag{1.2-23}$$

with $\omega_e L = X_L$, which is referred to as the inductive reactance. The phasor form of (1.2-23) is

$$\tilde{V}_L = jX_L\, \tilde{I}_L \tag{1.2-24}$$

Thus, the voltage across the inductor leads the current through it by $\pi/2$. That is, the current through the inductor lags the voltage across it by $\pi/2$ $\left[\theta_{esv}(0) = \theta_{esi}(0) + \frac{1}{2}\pi\right]$. This is shown in Fig. 1.2-1.

For the capacitor

$$V_C = \frac{1}{C}\int I_C\, dt$$
$$= \frac{1}{\omega_e C}\sqrt{2}I \cos\left[\omega_e t + \theta_{esi}(0) - \frac{1}{2}\pi\right] \tag{1.2-25}$$

Following the procedure used for the inductor, the phasor voltage across it becomes

$$\tilde{V}_C = -jX_C\tilde{I}_C \tag{1.2-26}$$

where $X_C = 1/\omega_e C$, the capacitive reactance. The voltage across the capacitor lags the current through it by $\pi/2$ $\left[\theta_{esv}(0) = \theta_{esi}(0) - \frac{1}{2}\pi\right]$, or the current through the capacitor leads the voltage across it by $\pi/2$. This is also shown in Fig. 1.2-1.

A series *RLC* circuit is shown in Fig. 1.2-2. From Fig. 1.2-2,

$$\tilde{V}_s = Z\tilde{I}_s \tag{1.2-27}$$

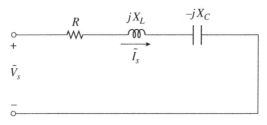

Figure 1.2-2 Phasor equivalent circuit for a series *RLC* circuit.

where $Z = R + j(X_L - X_C)$. We should be careful here. Some prefer to write (1.2-27) as $R + jX$ where X is $X_L + X_C$ and let X_C be negative. This is essentially a matter of choice and does not change the end result. We will deal primarily with X_L and not X_C, therefore, this will have little impact on our work; nevertheless, since some authors will use a negative X_C, we should make the reader aware of this difference.

It is appropriate to discuss the notation that will be used throughout the text. When an equation is written with the variables in lowercase letters, it is valid for transient and steady state. If the variables are written with uppercase letters, the equation is a function of time and valid for instantaneous steady-state conditions. Equation (1.2-27) is a phasor equation representing steady-state sinusoidal variables and are written in uppercase letters with an over tilde.

1.2.1 Power and Reactive Power

The instantaneous steady-state power is

$$
\begin{aligned}
P &= V_s I_s \\
&= \sqrt{2}V \cos[\omega_e t + \theta_{ev}(0)]\sqrt{2}I \cos[\omega_e t + \theta_{ei}(0)]
\end{aligned}
\tag{1.2-28}
$$

where V and I are rms values. After some manipulation, we can write (1.2-28) as

$$
P = VI \cos[\theta_{ev}(0) - \theta_{ei}(0)] + VI \cos[2\omega_e t + \theta_{ev}(0) + \theta_{ei}(0)]
\tag{1.2-29}
$$

The instantaneous steady-state power given by (1.2-29) varies about an average value at a frequency of $2\omega_e$. That is, the second term of (1.2-29) has a zero average value and the average power P_{ave} may be written as

$$
P_{ave} = |\tilde{V}_s| |\tilde{I}_s| \cos[\theta_{ev}(0) - \theta_{ei}(0)]
\tag{1.2-30}
$$

where $|\tilde{V}_s|$ and $|\tilde{I}_s|$ are V and I, respectively, which are the magnitudes of the phasors (rms value), $\theta_{ev}(0) - \theta_{ei}(0)$ is referred to as the *power factor angle* ϕ_{pf}, and $\cos[\theta_{ev}(0) - \theta_{ei}(0)]$ is the *power factor*. Power is in watts. If current is assumed positive in the direction of voltage drop, then (1.2-30) is positive if power is consumed and negative if power is generated. It is interesting to point out that in going from (1.2-28) to (1.2-29), the coefficient of the two right-hand terms is $1/2(\sqrt{2}V\sqrt{2}I)$ or one half the product of the peak values of the sinusoidal variables. Therefore, it was considered more convenient to use the rms values for the phasors, whereupon average steady-state power could be calculated by the product of the magnitude of the voltage and current phasors as given by (1.2-30).

The reactive power is defined as

$$
Q = |\tilde{V}_s| |\tilde{I}_s| \sin[\theta_{ev}(0) - \theta_{ei}(0)]
\tag{1.2-31}
$$

The units of Q are var (volt-ampere reactive). An inductance is said to absorb reactive power where the current lags the voltage by 90° and Q is positive. In the case of a capacitor, where the current leads the voltage by 90°, Q is supplied and is negative. Actually, Q is a measure of the interchange of energy supplied by the source that is stored in the electric (capacitor) and magnetic (inductor) fields. However, unlike instantaneous real power, the average value of instantaneous reactive power is zero.

We would like to minimize reactive power flow over the transmission lines in a power system. In other words, we would like to transmit only real power from the source to the load. The loads are generally inductive; therefore, capacitors are often placed in parallel with the load to interchange reactive power with the inductive load thus preventing the interchange current from flowing over the transmission line. This is often referred to as power factor correction since the transmission power factor approaches unity.

Example 1.A Phasor Analysis

The parameters of a series RLC circuit are $R = 6\,\Omega$, $L = 20$ mH, and $C = 1 \times 10^3\,\mu$F. The 60-Hz applied voltage is $V_s = 155.6 \cos \omega_e t$. Calculate $\tilde{I}_s$, P_{ave}, and Q and draw the phasor diagram. From the expression of V_s,

$$\tilde{V}_s = 110\underline{/0°}\,\text{V} \tag{1A-1}$$

Now, $\omega_e = 2\pi f = 2\pi \times 60 = 377$ rad/s and

$$\begin{aligned}
Z &= R + j(X_L - X_C) \\
&= R + j\left(\omega_e L - \frac{1}{\omega_e C}\right) \\
&= 6 + j\left(377 \times 20 \times 10^{-3} - \frac{1}{377 \times 1 \times 10^{-3}}\right) = 7.73\underline{/39.1°}\,\Omega
\end{aligned} \tag{1A-2}$$

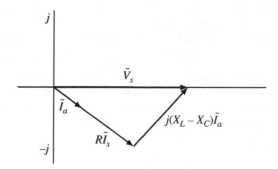

Figure 1.A-1 Phasor diagram.

$$\tilde{I}_s = \frac{\tilde{V}_s}{Z} = \frac{110\underline{/0°}}{7.73\underline{/39.1°}} = 14.2\underline{/-39.1°} \text{ A} \tag{1A-3}$$

$$P_{ave} = |\tilde{V}_s| \, |\tilde{I}_s| \cos\phi_{pf} \tag{1A-4}$$

where

$$\begin{aligned}\phi_{pf} &= \theta_{ev}(0) - \theta_{ei}(0) \\ &= 0 - (-39.1°) = 39.1°\end{aligned} \tag{1A-5}$$

$$\begin{aligned}P_{ave} &= 110 \times 14.2 \cos 39.1° \\ &= 1212.2 \text{ W}\end{aligned} \tag{1.A-6}$$

$$\begin{aligned}Q &= |\tilde{V}_s| \, |\tilde{I}_s| \sin\phi_{pf} \\ &= 110 \times 14.2 \sin 39.1° = 985.1 \text{ vars}\end{aligned} \tag{1.A-7}$$

The phasor diagram is shown Fig. 1.A-1.

SP1.2-1. Express the instantaneous steady-state power for Example 1.A. [Substitute into (1.2-29)].

SP1.2-2. Redraw the phasor diagram shown in Fig. 1.A-1 showing $jX_L\tilde{I}_s$ and $-jX_C\tilde{I}_s$ as individual voltages. [Show $jX_L\tilde{I}_s$ and then from the terminus of $jX_L\tilde{I}_s$, show $-jX_C\tilde{I}_s$].

SP1.2-3. We know that $P_{ave} = |\tilde{I}_s|^2 R$, does $Q = |\tilde{I}_s|^2 X_L - |\tilde{I}_s|^2 X_C$? [Yes]

SP1.2-4. If $\tilde{V}_s = 1\underline{/0°}$ V and $\tilde{I}_s = 1\underline{/180°}$ A in the direction of the voltage drop, calculate Z and P_{ave}. Is power generated or consumed? [$(-1+j0)$ ohms, 1 watt, generated]

SP1.2-5. Express the instantaneous power for 60-Hz voltage, $\tilde{V}_s = 1\underline{/0°}$, applied to a resistive circuit, $\tilde{I}_s = 1\underline{/0°}$. [$1 + \cos 754t$]

SP1.2-6. Repeat SP1.2-5 for (a) an inductance, $\tilde{I}_s = I_L\underline{/-90°}$ and (b) a capacitance, $\tilde{I}_s = I_C\underline{/90°}$. [(a) $I_L \cos(754t - 90°)$, (b) $I_C \cos(754t + 90°)$]

1.3 Stationary Magnetically Linear Systems

Before analyzing electromagnetic systems with motion, it is helpful to start with stationary electromagnetic systems. A stationary, single winding electromagnetic system is shown in Fig. 1.3-1. A coiled, conductive wire is referred to as a winding. Usually the wire is wound (coiled) around some structure called the core. Each loop of the winding around the core is referred to as a turn. The core is typically made up of ferromagnetic material to guide the magnetic flux created by current flowing in the wire. Magnetic flux prefers to travel through materials of high permeability, a property of ferromagnetic materials. Here, we use N to represent the number of turns of the winding.

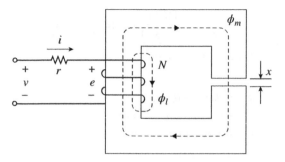

Figure 1.3-1 Single winding electromagnetic system.

In Fig. 1.3-1, ϕ_l is the leakage flux, which does not traverse the ferromagnetic core and ϕ_m is the magnetizing flux, which traverses the entire ferromagnetic core. Both ϕ_l and ϕ_m link N. The voltage equation is

$$v = ri + e \qquad (1.3\text{-}1)$$

where $e = p\lambda$ (p is the operator d/dt). The resistor voltage term is due to Ohm's law. The induced voltage term due to the change of flux linkages is Faraday's law.

From Fig. 1.3-1, the flux linking the winding is

$$\phi = \phi_l + \phi_m \qquad (1.3\text{-}2)$$

The magnetizing flux also travels across the slot in the core of width x. This slot is referred to as an air gap. In a structure such as this where the core is made of highly permeable material, the leakage flux, ϕ_l, is small and it generally makes up between 2 and 4% of the total flux linking the winding. Flux is said to link a winding if it travels through the turns of the winding. Both ϕ_l and ϕ_m link the winding. The total flux linked by the winding, called the flux linkage λ, is the flux through the winding multiplied by the number of turns of the winding

$$\begin{aligned} \lambda &= N\phi \\ &= N(\phi_l + \phi_m) \end{aligned} \qquad (1.3\text{-}3)$$

Next, we define magnetic equivalent circuits. Magnetic equivalent circuits are a model for magnetic systems based on Maxwell's equations and ideas from electrical circuit models. In magnetic circuits, we think of flux as current, magnetomotive force (mmf) as voltage, and reluctance as resistance. Ohm's law for magnetic circuits becomes

$$\phi_l = \frac{Ni}{\mathcal{R}_l} \qquad (1.3\text{-}4)$$

$$\phi_m = \frac{Ni}{\mathcal{R}_m} \tag{1.3-5}$$

where $\mathcal{R}$ is the reluctance and Ni is the mmf. The reluctance of the leakage path, $\mathcal{R}_l$, is large since a significant part of the path is in air. The reluctance of the ferromagnetic core and air gap may be calculated as

$$\mathcal{R}_m = \mathcal{R}_i + \mathcal{R}_g$$
$$= \frac{l_i}{\mu_r \mu_0 A_i} + \frac{x}{\mu_0 A_g} \tag{1.3-6}$$

where $l_i(x)$ is the length of the iron path (gap) and $A_i(A_g)$ is the cross-sectional area of the core (gap), and μ_0 is the permeability of free space ($4\pi \times 10^{-7}$ Wb/amp · m or H/m) and μ_r is the relative permeability. For air $\mu_r = 1$, for the ferromagnetic core, μ_r can be in the thousands, thus $\mathcal{R}_i < \mathcal{R}_g$. The equivalent magnetic circuit is shown in Fig. 1.3-2.

Substituting (1.3-4) and (1.3-5) into (1.3-3) yields

$$\lambda = N \left(\frac{Ni}{\mathcal{R}_l} + \frac{Ni}{\mathcal{R}_m} \right)$$
$$= \left(\frac{N^2}{\mathcal{R}_l} + \frac{N^2}{\mathcal{R}_m} \right) i \tag{1.3-7}$$
$$= (L_l + L_m) i$$

The self-inductance is $L_l + L_m$ where L_l and L_m are the leakage and magnetizing inductances, respectively. As seen above, inductance is defined as the relationship between current and flux linkage. The voltage equation given by (1.3-1) may now be written for a linear magnetic system as

$$v = ri + L \frac{di}{dt} \tag{1.3-8}$$

where

$$L = L_l + L_m \tag{1.3-9}$$

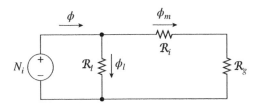

Figure 1.3-2 Magnetic equivalent circuit for the system shown in Fig. 1.3-1.

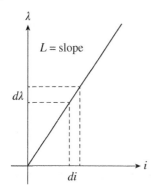

Figure 1.3-3 λi characteristic of a magnetically linear system.

Since we are assuming a linear magnetic system, core saturation and hysteresis are neglected and we have the λi plot shown in Fig. 1.3-3.

The magnetic equivalent circuit is similar to a resistive circuit. In particular, if we replace the magnetomotive force, mmf or Ni, with electromotive force, emf or voltage, and replace all reluctances, $\mathcal{R}$, with resistances R, then ϕ in Fig. 1.3-2 becomes the current i.

An important concept used in machine analysis is the idea of magnetic poles. The reader should have an intuition of north and south magnetic poles thanks to elementary physics classes and permanent magnets. Let us incorporate poles into our analysis of this stationary magnetically linear system. If fringing fields are neglected, the magnetizing flux, ϕ_m, travels uniformly across the air gap of Fig. 1.3-1. We can define magnetic north and south poles as the following: a north pole is a source of magnetic flux and a south pole is a sink for magnetic flux. To help determine pole assignment, place yourself on the member with the winding and where the positive flux enters the air gap is a north pole and where the positive flux enters the iron core is a south pole as shown in Fig. 1.3-4.

The concepts introduced in this chapter will be used throughout this text to analyze energy conversion systems. It is necessary to define quantities related to energy. The total energy stored in the field, W_f, may be expressed as

$$W_f = \int eidt$$
$$= \int id\lambda$$

(1.3-10)

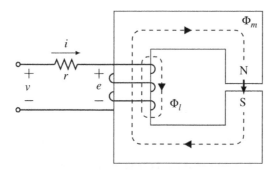

Figure 1.3-4 Repeat of Fig. 1.3-1 indicating north and south poles.

which is to the left of the plot shown in Fig. 1.3-3 for a given value of i. The area to the right of the plot is called the coenergy, W_c. It is expressed as

$$W_c = \int \lambda di \tag{1.3-11}$$

Coenergy is a quantity we define for analytical purposes. It is calculated from physical quantities, but has no direct physical meaning. Coenergy is convenient to formulate some expressions, for example, we can write from Fig. 1.3-3,

$$\lambda i = W_f + W_c \tag{1.3-12}$$

It should be clear that only for a magnetically linear system $W_f = W_c$.

1.3.1 Two-Winding Transformer

A two-winding transformer is shown in Fig.1.3-5. Here, we have mutual coupling between the two windings which we will take care of in a minute. Magnetic coupling refers to the situation where current through one winding creates flux which contributes to the flux linkage of another winding. Magnetic coupling is an essential aspect of transformers and electric machines. These devices may contain multiple windings that may be magnetically coupled. We analyze the two winding cases first. As always, we start with the winding voltage equations. The voltage equations are [1]

$$\begin{bmatrix} v_1 \\ v_2 \end{bmatrix} = \begin{bmatrix} r_1 & 0 \\ 0 & r_2 \end{bmatrix} \begin{bmatrix} i_1 \\ i_2 \end{bmatrix} + p \begin{bmatrix} \lambda_1 \\ \lambda_2 \end{bmatrix} \tag{1.3-13}$$

where the first term on the right-hand side comes from Ohm's law and the second from Faraday's law.

In Fig. 1.3-5, r_1 and r_2 are resistances of the windings and $e_1 = p\lambda_1$ and $e_2 = p\lambda_2$ where $p = \frac{d}{dt}$.

The flux linkages λ_1 and λ_2 may be expressed as

$$\begin{aligned} \lambda_1 &= N_1\phi_1 \\ &= N_1(\phi_{l1} + \phi_{m1} + \phi_{m2}) \\ &= \frac{N_1^2}{\mathcal{R}_{l1}}i_1 + \frac{N_1^2}{\mathcal{R}_m}i_1 + \frac{N_1N_2}{\mathcal{R}_m}i_2 \end{aligned} \tag{1.3-14}$$

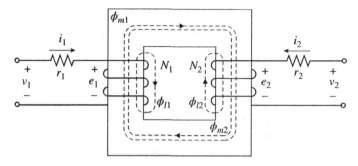

Figure 1.3-5 Two-winding transformer.

and

$$\begin{aligned}
\lambda_2 &= N_2 \phi_2 \\
&= N_2(\phi_{l2} + \phi_{m2} + \phi_{m1}) \\
&= \frac{N_2^2}{\mathcal{R}_{l2}} i_2 + \frac{N_2^2}{\mathcal{R}_m} i_2 + \frac{N_2 N_1}{\mathcal{R}_m} i_1
\end{aligned} \tag{1.3-15}$$

where $\phi_1(\phi_2)$ is the flux that flows through the winding with $N_1(N_2)$ turns. The self-inductance of the windings comes from the first two terms of (1.3-14) and (1.3-15). It would exist even if the other coil were not present. That is,

$$L_{11} = \frac{N_1^2}{\mathcal{R}_{l1}} + \frac{N_1^2}{\mathcal{R}_m} = L_{l1} + L_{m1} \tag{1.3-16}$$

For winding 2,

$$L_{22} = \frac{N_2^2}{\mathcal{R}_{l2}} + \frac{N_2^2}{\mathcal{R}_m} = L_{l2} + L_{m2} \tag{1.3-17}$$

It is clear that the self-inductances are independent of other windings. The coefficient of the last term of (1.3-14) and (1.3-15) is called the mutual inductance, that is,

$$L_{12} = L_{21} = \frac{N_1 N_2}{\mathcal{R}_m} \tag{1.3-18}$$

Therefore,

$$\begin{bmatrix} \lambda_1 \\ \lambda_2 \end{bmatrix} = \begin{bmatrix} L_{11} & L_{12} \\ L_{21} & L_{22} \end{bmatrix} \begin{bmatrix} i_1 \\ i_2 \end{bmatrix} \tag{1.3-19}$$

The mutual inductance which is new to most of us can be positive or negative depending on the relative direction of ϕ_{m1} and ϕ_{m2}. In this case it is positive; if, however, the sense of winding 2 or the current i_2 is reversed, the mutual inductance would be negative.

The inductances L_{m1}, L_{m2}, L_{12}, and L_{21} have a common term $\mathcal{R}_m$. This allows us to write the flux linkages in terms of L_{m1} or L_{m2} and a turns ratio. We do this to create an electric circuit model of the two-winding transformer. If we write λ_1 and λ_2 in terms of $L_{m1}(L_{m2})$, we are referring voltages, currents, and flux linkages to winding 1 (winding 2). Referring to winding 1, λ_1 becomes

$$\lambda_1 = L_{l1}i_1 + L_{m1}\left(i_1 + \frac{N_2}{N_1}i_2\right) \tag{1.3-20}$$

Substituting

$$i_2' = \frac{N_2}{N_1}i_2 \tag{1.3-21}$$

we see that i_2' flowing in N_1 produces the same mmf as i_2 flowing in N_2. Now, in order to make $v_2'i_2' = v_2i_2$,

$$v_2' = \frac{N_1}{N_2}v_2 \tag{1.3-22}$$

Now since λ_2 is in volt·sec, λ_2' becomes

$$\lambda_2' = \frac{N_1}{N_2}\lambda_2 \tag{1.3-23}$$

and λ_1 and λ_2' become

$$\lambda_1 = L_{l1}i_1 + L_{m1}\left(i_1 + i_2'\right) \tag{1.3-24}$$

$$\lambda_2' = L_{l2}'i_2' + L_{m1}\left(i_1 + i_2'\right) \tag{1.3-25}$$

The voltage equations for all variables referred to winding 1 are

$$\begin{bmatrix} v_1 \\ v_2' \end{bmatrix} = \begin{bmatrix} r_1 & 0 \\ 0 & r_2' \end{bmatrix}\begin{bmatrix} i_1 \\ i_2' \end{bmatrix} + p\begin{bmatrix} \lambda_1 \\ \lambda_2' \end{bmatrix} \tag{1.3-26}$$

where

$$L_{l2}' = \left(\frac{N_1}{N_2}\right)^2 L_{l2} \tag{1.3-27}$$

$$r_2' = \left(\frac{N_1}{N_2}\right)^2 r_2 \tag{1.3-28}$$

These equations suggest the equivalent circuit given in Fig. 1.3-6.

With two windings, the total energy stored in the fields becomes

$$W_f = \int (e_1i_1 + e_2i_2)\, dt \tag{1.3-29}$$

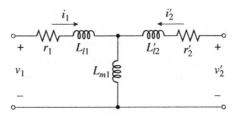

Figure 1.3-6 Transformer equivalent T circuit with winding 1 selected as reference winding.

or in terms of referred variables

$$W_f = \int \left(e_1 i_1 + e_2' i_2' \right) dt$$

$$= \int \left(i_1 d\lambda_i + i_2' d\lambda_2' \right)$$

(1.3-30)

where the 2 variables are referred to N_1.

Now,

$$d\lambda_1 = L_{11} di_1 + L_{12}' di_2'$$ (1.3-31)

$$d\lambda_2' = L_{12}' di_1 + L_{22}' di_2'$$ (1.3-32)

where

$$L_{12}' = \frac{N_1}{N_2} L_{12}$$ (1.3-33)

$$L_{22}' = L_{l2}' + \left(\frac{N_1}{N_2} \right)^2 L_{m2}$$ (1.3-34)

we can evaluate (1.3-30) in two steps; first we will hold i_2' at zero, thus $di_2' = 0$, and allow i_1 to go from zero to i_1. Thus,

$$W_{f(1)} = \int_0^{i_1} L_{11} \xi d\xi = \frac{1}{2} L_{11} i_1^2$$ (1.3-35)

where ξ is the dummy variable of integration. For the second step, we will hold i_1 at i_1 with $di_1 = 0$, and allow i_2' to go from zero to i_2'. Thus,

$$W_{f(2)} = \int_0^{i_2'} \left(L_{12}' i_1 d\xi + L_{22}' \xi d\xi \right)$$ (1.3-36)

The stored energy in the fields is

$$W_f = W_{f(1)} + W_{f(2)}$$

$$= \frac{1}{2} L_{11} i_1^2 + L_{12}' i_1 i_2' + \frac{1}{2} L_{22}' i_2'^2$$

(1.3-37)

It is clear that (1.3-37) includes the energy stored in the leakage inductances, which do not couple other fields also (1.3-37) is valid with or without the primes. For multi-winding systems, (1.3-12) becomes

$$\sum_{j=1}^{J} i_j \lambda_j = W_f + W_c \tag{1.3-38}$$

Example 1.B Parameters of the Transformer Equivalent Circuit

It is instructive to illustrate the method of deriving an equivalent T circuit from open- and short-circuit measurements of the transformer. When winding 2 of the two-winding transformer shown in Fig 1.3-6 is open circuited and a 60-Hz voltage of 110 V (rms) is applied to winding 1, the average power supplied to winding 1 is 6.66 W. The measured current in winding 1 is 1.05 A (rms). Next, with winding 2 short-circuited, the current flowing in winding 1 is 2 A when the applied 60-Hz voltage is 30 V (rms). The average input power is 44 W. If we assume $L_{l1} = L'_{l2}$, an approximate equivalent T circuit can be determined from these measurements with winding 1 selected as the reference winding.

The average power supplied to winding 1 may be expressed from (1.2-30) as

$$P_1 = \left|\tilde{V}_1\right| \left|\tilde{I}_1\right| \cos \varphi_{pf} \tag{1B-1}$$

where

$$\phi_{pf} = \theta_{ev}(0) - \theta_{ei}(0) \tag{1B-2}$$

Here, $\tilde{V}_1$ and $\tilde{I}_1$ are phasors with the positive direction of $\tilde{I}_1$ taken in the direction of voltage drop, and $\theta_{ev}(0)$ and $\theta_{ei}(0)$ are the phase angles of $\tilde{V}_1$ and $\tilde{I}_1$, respectively. Solving for ϕ_{pf} during the open-circuit test, we have

$$\phi_{pf} = \cos^{-1}\frac{P_1}{\left|\tilde{V}_1\right|\left|\tilde{I}_1\right|} = \cos^{-1}\frac{6.66}{(110)(1.05)} = 86.7° \tag{1B-3}$$

Although $\phi_{pf} = -86.7°$ is also a legitimate solution of (1B-3), the positive value is taken since $\tilde{V}_1$ leads $\tilde{I}_1$ in an inductive circuit. With winding 2 open-circuited, the input impedance of winding 1 is

$$Z = \frac{\tilde{V}_1}{\tilde{I}_1} = r_1 + j(X_{l1} + X_{m1}) \tag{1B-4}$$

With $\tilde{V}_1$ as the reference phasor, $\tilde{V}_1 = 110\underline{/0°}$, $\tilde{I}_1 = 1.05\underline{/-86.7°}$. Thus,

$$r_1 + j(X_{l1} + X_{m1}) = \frac{110\underline{/0°}}{1.05\underline{/-86.7°}} = 6 + j104.6\,\Omega \tag{1B-5}$$

From (1B-5), $r_1 = 6\,\Omega$. We also see from (1B-5) that $X_{l1} + X_{m1} = 104.6\,\Omega$.

For the short-circuit test, we will assume that $\tilde{I}_1 = -\tilde{I}_2'$ since transformers are designed so that at rated frequency $X_{m1} \gg |r_2' + jX_{l2}'|$. Hence, using (1B-1) again,

$$\phi_{pf} = \cos^{-1}\frac{44}{(30)(2)} = 42.8° \tag{1B-6}$$

In this case, the input impedance is $Z = (r_1 + r_2') + j(X_{l1} + X_{l2}')$. This may be determined as

$$Z = \frac{30/0°}{2/-42.8°} = 11 + j10.2\,\Omega \tag{1B-7}$$

Hence, $r_2' = 11 - r_1 = 5\,\Omega$ and, since it is assumed that $X_{l1} = X_{l2}'$, both are $10.2/2 = 5.1\,\Omega$. Therefore, $X_{m1} = 104.6 - 5.1 = 99.5\,\Omega$. In summary, $r_1 = 6\,\Omega$, $L_{l1} = 13.5\,\text{mH}$, $L_{m1} = 263.9\,\text{mH}$, $r_2' = 5\,\Omega$, and $L_{l2}' = 13.5\,\text{mH}$. It is left to the reader to verify the conversion from X's to L's.

SP1.3-1. Show that the total field energy if a third winding is added to Fig. 1.3-5 is

$$W_f = \frac{1}{2}L_{11}i_1^2 + \frac{1}{2}L_{22}i_2^2 + \frac{1}{2}L_{33}i_3^2 + L_{12}i_1i_2 + L_{13}i_1i_3 + L_{23}i_2i_3$$

SP1.3-2. Draw the equivalent circuit for a three-winding transformer with all variables referred to winding 1.

SP1.3-3. Consider the transformer and parameters calculated in Example 1.B. Winding 2 is short-circuited and 12 V (dc) is applied to winding 1. Calculate the steady-state values of i_1 and i_2. Repeat with winding 2 open-circuited. [$I_1 = 2\,\text{A}$ and $I_2 = 0$ in both cases]

1.4 Winding Configurations

The previous sections analyzed basic electromagnetic structures. Now, we will introduce more complicated geometries useful for the construction of electric machines. Electric machines are configurations of windings and ferromagnetic material that create and guide magnetic fields. The magnetic fields interact to create forces that turn the rotor.

The stator windings are shown in Fig. 1.4-1. We can see this multiphase stator is very involved. In this section, we are going to consider the stator windings as simply as possible.

Figure 1.4-1   Stator windings of a multiphase machine.

To form the stator of an electric machine, conductive wire is wound in the slots of a steel structure. The number of turns or coils of the stator windings of most ac machines are distributed to approximate a space sinusoid as shown in Fig. 1.4-2. In Section 1.2, we used the "*s*" subscript to denote sinusoidal variables. In Fig. 1.4-2, we use the "*s*" subscript to denote stator or stationary. We will use this definition of *s* as a subscript or superscript for the remainder of the text. Also in Fig. 1.4-2, the "*as*" subscript denotes the variables associated with the *a*-phase of the stator. In some machines, great pains are taken to obtain a sinusoidal distribution of the stator windings to meet harmonic specifications. We attempt to distribute windings sinusoidally because sinusoidal currents through sinusoidally distributed windings create a constant amplitude rotating air-gap mmf. We use the terms rotating air-gap mmf and rotating magnetic field interchangeably. We will talk about the rotating air-gap mmf in detail in Chapter 2. With a constant amplitude rotating air-gap mmf, a constant power or torque is produced.

In Fig. 1.4-2, each winding segment $as_1 - as_1'$, $as_2 - as_2'$, $as_3 - as_3'$, and $as_4 - as_4'$ has nc_s coils for each $\otimes$ or $\odot$. Positive current is into the paper indicated by $\otimes$ and out of the paper at $\odot$. The current through the windings is alternating so the cross, $\otimes$, and $\odot$, will change; however, we are looking at an instant of time where positive current is in at as_1, as_2, as_3, and as_4.

If we follow the path of assumed positive current i_{as} flowing in the *as* winding, we see that current enters as_1, depicted by $\otimes$, to indicate that the assumed direction of positive current is down the length of the stator in an axial direction (into the paper). Current flows down the length of the stator, loops at the end, and flows

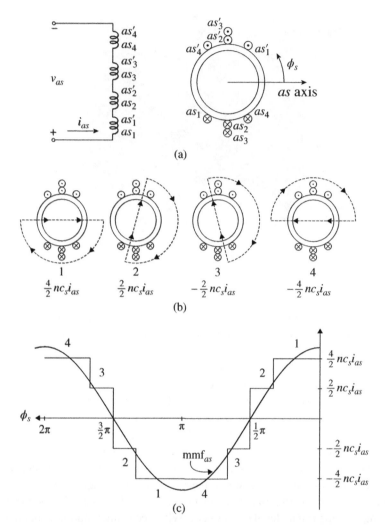

Figure 1.4-2 Elementary sinusoidally distributed windings. (a) Winding connections and distribution, (b) Ampere's law, and (c) mmf$_{as}$.

back down the length of the stator and out at as'_1, depicted by $\odot$. Note that as_1 and as'_1 are placed in stator slots that span π radians. This is referred to as the *winding pitch* of π radians which is characteristic of a two-pole machine. Now, as_1 around to as'_1 is referred to as a *coil* and as_1 or as'_1 is a *coil side*. In practice, a coil will contain more than one conductor. Current flows into as_1 in a conductor and out of as'_1 via the same conductor. The conductor, which is insulated, may then be looped back to as_1 and the winding of the conductor around the $as_1 - as'_1$ path repeated, thereby forming a coil with numerous turns. The number of conductors in a coil side tells us the number of turns in the coil, which is denoted as nc_s.

Once we have wound nc_s turns in the $as_1 - as_1'$ coil, we will take the same conductor and repeat this winding process to form the $as_2 - as_2'$ coil. We will assume that the same number of turns (nc_s) make up the $as_2 - as_2'$ coil as the $as_1 - as_1'$ coil and, similarly, for $as_3 - as_3'$ and $as_4 - as_4'$. We could have wound a different number of turns in each coil but we will assume that this was not done. Once the winding is wound, we can use the right-hand rule to give a meaning to the as axis shown in Fig. 1.4-2a. It is, by definition, the principal direction of the magnetic flux established by the assumed positive current flowing in the as winding. It is said to indicate the assumed positive direction of the magnetic axis of the as winding of this elementary sinusoidally distributed winding. The positive direction of the as axis reverses when i_{as} reverses.

Before getting into the mmf due to the current flowing in the winding let us consider the self-inductance of the winding. The rotor is round and the magnetizing flux established by this winding must cross the air gap twice and for positive current as shown in Fig. 1.4-2, the positive as axis is to the right. Since the air gap is uniform, the self-inductance is constant independent of rotor position. Therefore, the self-inductance is of the same form as given by (1.3-9). The difference is the leakage inductance makes up 5–15% of the self-inductance and the reluctance to the flux is dominated by the air gaps.

Ampere's law is

$$\oint \overline{H} \cdot dL = i \tag{1.4-1}$$

which says that the closed line integral of the mmf drops equals that current enclosed. For the instant shown in Fig. 1.4-2

$$\mathrm{mmf}(0) + \mathrm{mmf}(\pi) = N_s i_{as} \tag{1.4-2}$$

where one half of the mmf is dropped at $\phi_s = 0$ and one half at $\phi_s = \pi$. Since there is no point source of mmf, we will assume that rotor to stator is positive. Since the reluctance of air is much larger than iron, we will neglect the mmf drop in the iron and assume that the air gap is uniform, thus,

$$\mathrm{mmf}(0) = \frac{N_s}{2} i_{as} \tag{1.4-3}$$

$$\mathrm{mmf}(\pi) = -\frac{N_s}{2} i_{as} \tag{1.4-4}$$

Also the path of integration in Fig. 1.4-2 is 180°. Regardless of the type of rotor, the air gap is the same every 180° for the two-pole device. Following this same procedure for paths 2 through 4, we obtain the stepped plot of mmf_{as} as shown in Fig. 1.4-2c. The fundamental component of this stepped mmf is

$$\text{mmf}_{as} = \frac{N_s}{2} i_{as} \cos \phi_s \tag{1.4-5}$$

where N_s is the amplitude of the fundamental component of the Fourier transform of the winding distribution. For the winding distribution given in Fig. 1.4-2, N_s is $2.37\, nc_s$. Note, also that the sinusoidal distributed winding is denoted with $\otimes$ and $\odot$ placed at the maximum winding density.

The winding shown in Fig. 1.4-2 is an approximation of a sinusoidal distributed winding. In the case of a large generator, pains would be taken to distribute the winding much closer to a sinusoidal winding to minimize the voltage harmonics. Nevertheless, Ampere's law would be performed the same. Also, for multiphase machines, windings of two- or three-phase may exist in the same slot.

Example 1.C Air-Gap mmf for a Uniformly Distributed Winding

Consider the uniformly distributed winding arrangement shown in Fig. 1.C-1. Each coil has nc_s turns and the current in each turn is i_{as} with the positive direction as shown. Follow the procedure used in Fig. 1.4-2 to establish the air-gap mmf.

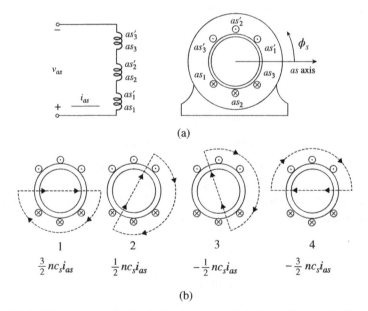

Figure 1.C-1 Elementary two-pole single-phase stator winding uniformly distributed. (a) Winding connections and distribution, (b) Ampere's law, and (c) mmf_{as}, here $N_s = 1.534\, nc_s$.

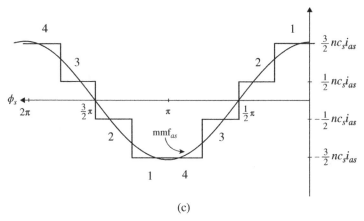

(c)

Figure 1.C-1 (Continued)

SP1.4-1. Express the sinusoidal approximation of mmf$_{as}$ if the assumed positive direction of i_{as} is reversed in Fig. 1.4-2. [mmf$_{as}$ = $-(1.4-5)$]

SP1.4-2. Assume that only the as_2 winding exists in Fig. 1.C-1a which has one turn. Sketch the air-gap mmf due to current i_1 flowing in this winding. [$\frac{1}{2}i_1$ for $-\frac{1}{2}\pi < \phi_s < \frac{1}{2}\pi$ and $-\frac{1}{2}i_1$ for $\frac{1}{2}\pi < \phi_s < \frac{2}{3}\pi$]

1.5 Two- and Three-Phase Stators

High-voltage transmission, most inverter-supplied electric drives, and the alternator of your car are examples of three-phase systems. Although two-phase systems are not common, a two-phase system is far less involved when it comes to machine analysis than its three-phase big sister. Fortunately, once the derivations have been set forth for a two-phase machine, the extension to a three-phase machine is straightforward and easily achieved. This section is devoted to the introduction of these multiphase systems.

By definition, a two-phase set of variables is balanced if the variables are equal-amplitude sinusoidal quantities in time quadrature (90° out of time phase). A three-phase set of variables is balanced if the sinusoidal variables are equal-amplitude quantities that are 120° out of time phase with each other.

1.5.1 Two-Phase Stator

In the broadest sense of the above definition, two-phase balanced sets may be expressed as

$$f_a(t) = \pm f \cos\theta_{ef} \tag{1.5-1}$$

$$f_b(t) = \pm f \sin\theta_{ef} \tag{1.5-2}$$

where

$$\theta_{ef}(t) = \int_0^t \omega_e(\xi)d\xi + \theta_{ef}(0) \tag{1.5-3}$$

In (1.5-1) and (1.5-2), $f_a(t)$ is the a-phase and $f_b(t)$ is the b-phase of voltage, current, or flux linkage. The amplitude f is assumed to be constant. In (1.5-3), ω_e is the electrical angular velocity and ξ is a dummy variable of integration. Equations (1.5-1) and (1.5-2) express four balanced two-phase sets. Like signs of (1.5-1) and (1.5-2) define balanced sets where $f_a(t)$ leads $f_b(t)$ by 90°, an ab sequence; for unlike signs $f_a(t)$ lags $f_b(t)$ by 90°, a ba sequence.

For steady-state balanced conditions, ω_e is constant and (1.5-3) becomes

$$\theta_{ef}(t) = \omega_e t + \theta_{ef}(0) \tag{1.5-4}$$

Whereupon (1.5-1) and (1.5-2) are written as

$$F_a(t) = \pm \sqrt{2}F \cos\left[\omega_e t + \theta_{ef}(0)\right] \tag{1.5-5}$$

$$F_b(t) = \pm \sqrt{2}F \sin\left[\omega_e t + \theta_{ef}(0)\right] \tag{1.5-6}$$

For like signs of (1.5-5) and (1.5-6), $\tilde{F}_a = j\tilde{F}_b$; for unlike signs, $\tilde{F}_a = -j\tilde{F}_b$.

Most large horsepower electric machines are three-phase and smaller household machines are two-phase machines powered from a single-phase source like a common wall outlet. For single-phase machines, a capacitor is generally connected in series with one of the windings (see Chapter 7). It is helpful to take a brief look at the stator winding arrangement of the two-phase machine, shown in Fig. 1.5-1. The windings are assumed to be identical in parameters and distribution. The displacement around the stator is denoted ϕ_s. The two windings are displaced 90° degrees from each other. The voltage equations may be written as

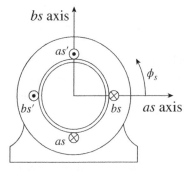

$$v_{as} = r_s i_{as} + \frac{d\lambda_{as}}{dt} \tag{1.5-7}$$

$$v_{bs} = r_s i_{bs} + \frac{d\lambda_{bs}}{dt} \tag{1.5-8}$$

where the subscripts as and bs denote phase a of the stator and phase b of the stator, respectively. In matrix form

$$\mathbf{v}_{abs} = \mathbf{r}_s \mathbf{i}_{abs} + p\boldsymbol{\lambda}_{abs} \tag{1.5-9}$$

Figure 1.5-1 Elementary two-pole two-phase sinusoidally distributed stator windings.

The stator has identical, sinusoidally distributed windings and the air gap is uniform.

The magnetic axes are orthogonal (thus $L_{asbs} = 0$) and the flux linkage equations may be written as

$$\lambda_{as} = L_{asas}i_{as}$$
$$= (L_{ls} + L_{ms})i_{as} \tag{1.5-10}$$

$$\lambda_{bs} = L_{bsbs}i_{bs}$$
$$= (L_{ls} + L_{ms})i_{bs} \tag{1.5-11}$$

where L_{ls} is the leakage inductance and L_{ms} is the magnetizing inductance of the stator windings. In matrix form

$$\begin{bmatrix} \lambda_{as} \\ \lambda_{bs} \end{bmatrix} = \begin{bmatrix} L_{ss} & 0 \\ 0 & L_{ss} \end{bmatrix} \begin{bmatrix} i_{as} \\ i_{bs} \end{bmatrix} \tag{1.5-12}$$

or

$$\boldsymbol{\lambda}_{abs} = \mathbf{L}_s \mathbf{i}_{abs} \tag{1.5-13}$$

where

$$\mathbf{L}_s = \begin{bmatrix} L_{ss} & 0 \\ 0 & L_{ss} \end{bmatrix} \tag{1.5-14}$$

and

$$L_{ss} = L_{ls} + L_{ms} \tag{1.5-15}$$

An important feature of multiphase systems is that the instantaneous power is constant for balanced operation. You are asked to show this in SP1.5-2. Recall that in a single-phase system the instantaneous power has an average value and a double frequency component.

1.5.2 Three-Phase Stator

The three-phase stator is shown in Fig. 1.5-2. A three-phase balanced set may be expressed as

$$f_a(t) = f \cos \theta_{ef} \tag{1.5-16}$$

$$f_b(t) = f \cos \left(\theta_{ef} - \frac{2}{3}\pi \right) \tag{1.5-17}$$

$$f_c(t) = f \cos \left(\theta_{ef} + \frac{2}{3}\pi \right) \tag{1.5-18}$$

where θ_{ef} is given by (1.5-3). This set is referred to as an abc sequence, since $f_a(t)$ leads $f_b(t)$ by 120° and $f_b(t)$ leads $f_c(t)$ by 120°. An acb sequence is obtained by interchanging $f_b(t)$ and $f_c(t)$, that is,

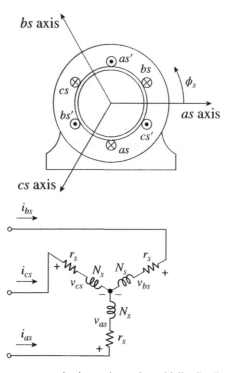

Figure 1.5-2 Elementary two-pole three-phase sinusoidally distributed stator windings.

$$f_a(t) = f \cos \theta_{ef} \tag{1.5-19}$$

$$f_b(t) = f \cos \left(\theta_{ef} + \frac{2}{3}\pi \right) \tag{1.5-20}$$

$$f_c(t) = f \cos \left(\theta_{ef} - \frac{2}{3}\pi \right) \tag{1.5-21}$$

For steady-state balanced conditions, the *abc* sequence may be written as

$$F_a(t) = \sqrt{2}F \cos \left[\omega_e t + \theta_{ef}(0) \right] \tag{1.5-22}$$

$$F_b(t) = \sqrt{2}F \cos \left[\omega_e t - \frac{2}{3}\pi + \theta_{ef}(0) \right] \tag{1.5-23}$$

$$F_c(t) = \sqrt{2}F \cos \left[\omega_e t + \frac{2}{3}\pi + \theta_{ef}(0) \right] \tag{1.5-24}$$

with $\tilde{F}_a = F \underline{/\theta_{ef}(0)}$, $\tilde{F}_b = F \underline{/\theta_{ef}(0) - \frac{2}{3}\pi}$, and $\tilde{F}_c = F \underline{/\theta_{ef}(0) + \frac{2}{3}\pi}$. For an *acb* sequence, $\tilde{F}_b$ and $\tilde{F}_c$ are interchanged.

A three-phase stator is shown in Fig. 1.5-2. Again, the stator has identical, sinusoidally distributed windings and the air gap is uniform. The magnetic axes of the windings are displaced 120° and the windings are often "wye-connected" as shown. The line-to-neutral voltage equations may be written as

$$v_{as} = r_s i_{as} + \frac{d\lambda_{as}}{dt} \tag{1.5-25}$$

$$v_{bs} = r_s i_{bs} + \frac{d\lambda_{bs}}{dt} \tag{1.5-26}$$

$$v_{cs} = r_s i_{cs} + \frac{d\lambda_{cs}}{dt} \tag{1.5-27}$$

where subscripts *as*, *bs*, and *cs* denote the three phases of the stator. In matrix form

$$\mathbf{v}_{abcs} = \mathbf{r}_s \mathbf{i}_{abcs} + p\boldsymbol{\lambda}_{abcs} \tag{1.5-28}$$

Since the windings are displaced 120° from each other, there is a mutual coupling between the stator windings. Let us assume that we can move the *bs* winding clockwise through the iron until it is "on top" of the *as* winding at $\phi_s = 0$. The coupling would be maximum positive. Now, assume we can rotate the *bs* winding counterclockwise back to $\phi_s = 120°$ where the mutual inductance between the *as* and *bs* windings can be approximated as

$$
\begin{aligned}
L_{asbs} &= L_{ms} \cos 120° \\
&= -\frac{1}{2} L_{ms}
\end{aligned}
\tag{1.5-29}
$$

where L_{ms} is the magnetizing inductance of the stator windings. Following this same approach, we can express the flux-linkage matrix as

$$
\begin{bmatrix} \lambda_{as} \\ \lambda_{bs} \\ \lambda_{cs} \end{bmatrix}
=
\begin{bmatrix}
L_{ss} & -\frac{1}{2}L_{ms} & -\frac{1}{2}L_{ms} \\
-\frac{1}{2}L_{ms} & L_{ss} & -\frac{1}{2}L_{ms} \\
-\frac{1}{2}L_{ms} & -\frac{1}{2}L_{ms} & L_{ss}
\end{bmatrix}
\begin{bmatrix} i_{as} \\ i_{bs} \\ i_{cs} \end{bmatrix}
\tag{1.5-30}
$$

where

$$L_{ss} = L_{ls} + L_{ms} \tag{1.5-31}$$

Equation (1.5-30) may also be written as

$$\boldsymbol{\lambda}_{abcs} = \mathbf{L}_s \mathbf{i}_{abcs} \tag{1.5-32}$$

1.5.3 Line-to-Line Voltage

In the case of a three-phase stator as shown in Fig. 1.5-2, the voltage rating is generally given as line-to-line voltage. For example, $\tilde{V}_{ab}$ is

$$\tilde{V}_{ab} = \tilde{V}_{as} - \tilde{V}_{bs} \tag{1.5-33}$$

For an *abc* sequence

$$
\begin{aligned}
\tilde{V}_{ab} &= V_s \underline{/0°} - V_s \underline{/-120°} \\
&= V_s(1 + j0) - V_s(-0.5 - j0.866) \\
&= \sqrt{3} V_s \underline{/30°}
\end{aligned} \tag{1.5-34}
$$

$$
\begin{aligned}
\tilde{V}_{bc} &= V_s \underline{/-120°} - V_s \underline{/120°} \\
&= V_s(-0.5 - j0.866) - V_s(-0.5 + j0.866) \\
&= \sqrt{3} V_s \underline{/-90°}
\end{aligned} \tag{1.5-35}
$$

$$
\begin{aligned}
\tilde{V}_{ca} &= V_s \underline{/120°} - V_s \underline{/0°} \\
&= V_s(-0.5 + j0.866) - V_s(1 + j0) \\
&= \sqrt{3} V_s \underline{/150°}
\end{aligned} \tag{1.5-36}
$$

The magnitude of the line-to-line voltages is $\sqrt{3}$ times the phase voltages and shifted 30° ccw.

Example 1.D Voltage Equations for a Three-Wire System

A three-phase stator similar to that given in Fig. 1.5-2 is connected to a three-phase source as shown in Fig. 1.D-1. Assume the stator is symmetrical, that is, the windings have the same resistance and same number of turns and displaced 120°. The stator configuration could be that of induction or synchronous machine. The

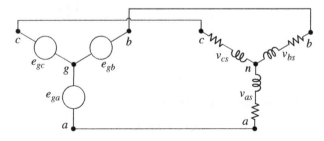

Figure 1.D-1 Three-phase source connected to symmetrical stator windings.

source voltages e_{ga}, e_{gb}, and e_{gc} may be of any form. Express v_{as}, v_{bs}, and v_{cs} in terms of e_{ga}, e_{gb}, and e_{gc}.

From Fig. 1.D-1, we can write

$$e_{ga} = v_{as} + v_{ng} \tag{1D-1}$$

$$e_{gb} = v_{bs} + v_{ng} \tag{1D-2}$$

$$e_{gc} = v_{cs} + v_{ng} \tag{1D-3}$$

Adding (1D-1) through (1D-3) yields

$$e_{ga} + e_{gb} + e_{gc} = v_{as} + v_{bs} + v_{cs} + 3v_{ng} \tag{1D-4}$$

Let us look at $v_{as} + v_{bs} + v_{cs}$. From (1.5-25) through (1.5-27)

$$v_{as} + v_{bs} + v_{cs} = r_s(i_{as} + i_{bs} + i_{cs}) + p(\lambda_{as} + \lambda_{bs} + \lambda_{cs}) \tag{1D-5}$$

In a three-wire, wye-connected stator, the sum of $i_{as} + i_{bs} + i_{cs}$ must be zero regardless of the form of the currents. Now from (1.5-30)

$$\lambda_{as} + \lambda_{bs} + \lambda_{cs} = L_{ss}(i_{as} + i_{bs} + i_{cs}) - L_{ms}(i_{as} + i_{bs} + i_{cs}) = 0 \tag{1D-6}$$

Thus,

$$v_{as} + v_{bs} + v_{cs} = 0 \tag{1D-7}$$

Will this be the case when we bring the rotor into play? We will find that for the electromechanical devices we will consider, it will be true. Substituting (1D-7) into (1D-4) yields

$$v_{ng} = \frac{1}{3}\left(e_{ga} + e_{gb} + e_{gc}\right) \tag{1D-8}$$

Going back to (1D-1) through (1D-3), we can write

$$\begin{aligned} v_{as} &= e_{ga} - v_{ng} \\ &= \frac{2}{3}e_{ga} - \frac{1}{3}\left(e_{gb} + e_{gc}\right) \end{aligned} \tag{1D-9}$$

$$\begin{aligned} v_{bs} &= e_{gb} - v_{ng} \\ &= \frac{2}{3}e_{gb} - \frac{1}{3}\left(e_{gc} + e_{ga}\right) \end{aligned} \tag{1D-10}$$

$$\begin{aligned} v_{cs} &= e_{gc} - v_{ng} \\ &= \frac{2}{3}e_{gc} - \frac{1}{3}\left(e_{ga} + e_{gb}\right) \end{aligned} \tag{1D-11}$$

We will make use of these equations when considering electric drives.

SP1.5-1. In Fig. 1.D-1, let $e_{ga} = 1$, $e_{gb} = 0$, and $e_{gc} = \cos \omega_e t$. Determine v_{as}, v_{bs}, and v_{cs}. $\left[\frac{2}{3} - \frac{1}{3}\cos\omega_e t; \quad -\frac{1}{3} - \frac{1}{3}\cos\omega_e t; \quad -\frac{1}{3} + \frac{2}{3}\cos\omega_e t\right]$. Note that $v_{as} + v_{bs} + v_{cs} = 0$.

SP1.5-2. In a two-phase system, let $V_a = \sqrt{2}V \cos[\omega_e t + \theta_{ev}(0)]$, $I_a = \sqrt{2}I \cos[\omega_e t + \theta_{ei}(0)]$, $V_b = \sqrt{2}V \sin[\omega_e t + \theta_{ev}(0)]$, and $I_b = \sqrt{2}I \sin[\omega_e t + \theta_{ei}(0)]$. Show that the total instantaneous power is $P = 2VI \cos[\theta_{ev}(0) - \theta_{ei}(0)]$.

SP1.5-3. Express the line-to-line voltages for an *acb* sequence. $[\tilde{V}_{ab} = \sqrt{3}V_s\underline{/-30°}, \tilde{V}_{bc} = \sqrt{3}V_s\underline{/90°}, \tilde{V}_{ca} = \sqrt{3}V_s\underline{/-150°}]$

1.6 Problems

1 Derive (1.2-26).

2 Derive (1.3-25).

3 Consider Fig. 1.3-5. The negative terminal of winding 1 is connected to the positive terminal of winding 2 and 110 V (rms) is applied between the positive terminal of winding 1 to the negative terminal of winding 2. Express the input impedance.

4 During the open-circuit test performed in Example 1.B, the rms voltage across the open-circuit 2 winding was 34.8 V. Determine X_{m2} $(\omega_e L_{m2})$.

5 Show that (1.3-38) is true for the two-winding system given in Fig. 1.3-5.

6 Determine mmf for a winding distribution uniformly as shown in Fig. 1.6-1 where coils with nc_s turns are 30° apart.

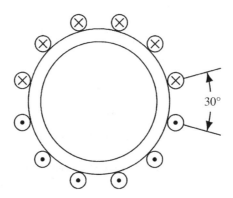

Figure 1.6-1 Uniform winding distribution.

7 Consider Example 1.D and Fig. 1.D-1. The load is symmetrical and the source voltages are given in Fig. 1.6-2. (a) Plot v_{ng}, v_{as}, v_{bs}, and v_{cs}. (b) Connect n to g in Fig. 1.D-1 and repeat part (a).

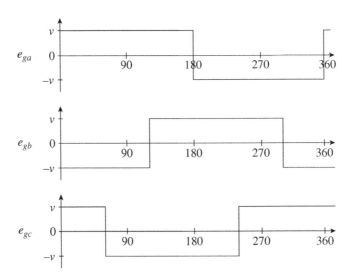

Figure 1.6-2 Waveforms of the source voltages of Fig. 1.D-1.

8 Assume that the direction of positive current is reversed in winding 2 of Fig. 1.3-5. Express (a) L_{12} in terms of N_1, N_2, and $\mathcal{R}_m$; (b) λ_1 and λ_2 in the form of (1.3-14) and (1.3-15); (c) λ_1 and λ_2' in the form of (1.3-24) and (1.3-25); and (d) v_1 and v_2' in the form of (1.3-26).

9 The parameters of a transformer are: $r_1 = r_2' = 10\,\Omega$, $L_{m1} = 300\,\text{mH}$, and $L_{l1} = L_{l2}' = 30\,\text{mH}$. A 10-V peak-to-peak 30-Hz sinusoidal voltage is applied to winding 1. Winding 2 is short-circuited. Assume $i_1 = -i_2'$. Calculate the phasor $\tilde{I}_1$ with $\tilde{V}_1$ at zero degrees.

10 A transformer with two windings has the following parameters: $r_1 = r_2 = 1\,\Omega$, $L_{m1} = 1\,\text{H}$, $L_{l1} = L_{l2} = 0.01\,\text{H}$, and $N_1 = N_2$. A $2 - \Omega$ load resistance R_L is connected across winding 2. $V_1 = 2\cos 400t$. (a) Calculate $\tilde{I}_1$. (b) Express I_1.

Reference

1 P. C. Krause, *Analysis of Electric Machinery*, McGraw-Hill Book Company, New York, 1986.

2

Analysis of the Symmetrical Stator

2.1 Introduction

Two important classes of ac electric machines are induction and synchronous machines. Both operate thanks to Tesla's rotating magnetic field. The induction machine is covered in Chapter 3. The synchronous machine is covered in Chapter 4. Although the rotors of induction and synchronous machines are different, the stators are essentially identical. The analysis of the symmetrical stator common to both is carried out in this chapter. It is called a symmetrical stator because each stator winding has the same distribution and electrical parameters.

This chapter also introduces a change of variables to analyze ac electric machines. It is a mathematical transformation that simplifies the analysis. The transformation is viewing Tesla's rotating magnetic field as an observer rotating at different speeds including zero, i.e. different frames of reference. In electric machine analysis, the transformation or "reference frame theory" is applied to physical quantities associated with the machine windings, in particular, voltages, currents, and flux linkages. The transformed versions of physical winding variables are the variables of "imaginary" or "fictitious" circuits that rotate at an arbitrary speed or, in other words, fictitious circuits that exist in a specific reference frame. Reference frame theory is the basis for modern analysis of ac electric machines.

2.2 Tesla's Rotating Magnetic Field

In the case of a single-phase stator winding, the sinusoidal approximation of the air-gap mmf is given by (1.4-5). Let us think about this for a minute: (1.4-5) is a product of i_{as} and $\cos\phi_s$, whereupon, the air-gap mmf will be nonexistent

Introduction to Modern Analysis of Electric Machines and Drives, First Edition.
Paul C. Krause and Thomas C. Krause.
© 2023 The Institute of Electrical and Electronics Engineers, Inc.
Published 2023 by John Wiley & Sons, Inc.

whenever $i_{as} = 0$. This is not too desirable since we will find that in order to produce a constant steady-state torque, we need a constant amplitude rotating air-gap mmf (north and south poles) produced by the stator. Is it possible to place another winding on the stator and arrange its position and the time sequence of the current flowing in it so that its mmf would complement the mmf due to the *as* winding resulting in a rotating air-gap mmf whose amplitude is constant? Yes, fortunately, and Tesla showed us how; the symmetrical two-pole two-phase stator is the most elementary example [1].

2.2.1 Two-Pole Two-Phase Stator

In the case of a symmetrical two-pole two-phase stator, the *bs* winding is located $\pi/2$ radians from the *as* winding and it is identical in distribution and parameters to the *as* winding. Let us assume it is positioned as illustrated in Fig. 2.2-1. Therein, a circle placed at the position of maximum turn density indicates a sinusoidally distributed winding. The mmf across one air gap created by i_{bs} is approximated as

$$\text{mmf}_{bs} = \frac{N_s}{2} i_{bs} \sin \phi_s \qquad (2.2\text{-}1)$$

where now N_s is the amplitude of the fundamental component of the winding distribution. Thus, since we are assuming a magnetically linear system, one half of the stator air-gap mmf due to the two-pole stator windings would be

$$\text{mmf}_s = \text{mmf}_{as} + \text{mmf}_{bs}$$

$$= \frac{N_s}{2} (i_{as} \cos \phi_s + i_{bs} \sin \phi_s) \qquad (2.2\text{-}2)$$

where the *s* subscript now refers to the stator not to be confused with *s* used in Section 1.2 for a sinusoidal variable. This relationship is perhaps the most

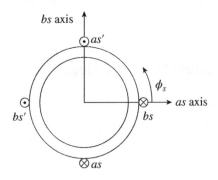

Figure 2.2-1 Elementary two-pole two-phase sinusoidally distributed stator windings.

important equation we will run into in this text since it is key to reference frame theory and the operation and analysis of ac machines.

A two-phase steady-state balanced set of stator currents may be expressed for an ab-sequence as

$$I_{as} = \sqrt{2}I_s \cos\left[\omega_e t + \theta_{esi}(0)\right] \tag{2.2-3}$$

$$I_{bs} = \sqrt{2}I_s \sin\left[\omega_e t + \theta_{esi}(0)\right] \tag{2.2-4}$$

where I_s is the rms value of the phase current, ω_e is the electrical angular velocity in rad/s, and $\theta_{esi}(0)$ is the phase angle of the currents. The stator currents are defined as positive into the machine. This is convenient for motor action where voltages applied to the stator windings cause current to flow and produce torque. For this balanced set, $\widetilde{I}_{bs} = -j\widetilde{I}_{as}$.

Since Tesla's magnetic field rotates relative to the stator, perhaps, looking at it from a rotating axis will help us visualize what is happening. In Fig. 2.2-2, we have added a third axis, the q axis, to Fig. 2.2-1. We will explain why we use "q" later. As shown in Fig. 2.2-2, the concept of a q axis and associated winding is no different from the as or bs axis, except that the winding and q axis can rotate at any arbitrary angular velocity ω or it can be stationary ($\omega = 0$). We are free to specify ω. The sinusoidally distributed winding with the q axis is fictitious except when $\theta = 0$ where it is the as winding. The as and bs axes are fixed with their associated stator windings and serve as a stationary reference. We are free to specify it. The angle from the as axis to the q axis is θ and for a constant ω is

$$\theta = \omega t + \theta(0) \tag{2.2-5}$$

The displacement ϕ shown in Fig. 2.2-2 is from the q axis just as ϕ_s is the displacement from the as axis. Let us relate a rotating position to an adjacent position on the stator. From Fig. 2.2-2,

$$\phi_s = \theta + \phi \tag{2.2-6}$$

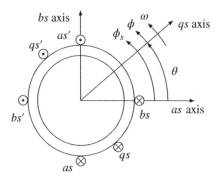

Figure 2.2-2 Elementary two-pole two-phase sinusoidally distributed stator windings with a third magnetic axis (q axis).

If we substitute (2.2.-6) into (2.2-2), we obtain

$$\text{mmf}_s = \frac{N_s}{2} \left[i_{as} \cos(\theta + \phi) + i_{bs} \sin(\theta + \phi) \right] \tag{2.2-7}$$

Substituting (2.2-3) and (2.2-4) into (2.2-7) for i_{as} and i_{bs}, respectively, and (2.2-5) for θ and after some work with trigonometric identities, the steady-state mmf_s becomes

$$\text{mmf}_s = \frac{N_s}{2} \sqrt{2} I_s \cos \left[(\omega_e - \omega)t + \theta_{esi}(0) - \theta(0) - \phi \right] \tag{2.2-8}$$

Recall that ω_e is the angular velocity of the electrical system and ω is the angular velocity of the q axis, $\theta(0)$ is the time-zero position of the q axis which will be considered to be zero unless otherwise specified, and ϕ is the displacement from the q axis. We have chosen to work with one of the two air gaps. The total mmf is two times (2.2-8).

If we let the angular velocity of the q axis to be zero ($\omega = 0$) and $\theta(0) = 0$, then $\phi = \phi_s$ and we would be viewing Tesla's rotating magnetic field as a stationary observer. In other words, the q-axis is stationary. In this case, (2.2-8) becomes

$$\text{mmf}_s^s = \frac{N_s}{2} \sqrt{2} I_s \cos \left[\omega_e t + \theta_{esi}(0) - \phi_s \right] \tag{2.2-9}$$

where we have added a superscript s to emphasize that we are observing Tesla's rotating magnetic field from a stationary frame of reference. Since both $\theta_{esi}(0)$ and ϕ_s are constants, mmf_s^s is a sinusoidal variation of frequency ω_e. If ϕ_s is zero, as we have mentioned, we would be positioned (fixed) at the as axis ($\phi_s = 0$ in Fig. 2.2-2) and the mmf_s^s would be pulsating at ω_e.

If we now let $\omega = \omega_e$, (2.2-8) becomes

$$\text{mmf}_s^e = \frac{N_s}{2} \sqrt{2} I_s \cos \left[\theta_{esi}(0) - \phi_e \right] \tag{2.2-10}$$

where ϕ_e (a renamed version of ϕ) is the displacement from the q axis which is rotating at ω_e. Since $\theta_{esi}(0)$ is constant for balanced steady-state conditions and let us say we have 360° vision, then (2.2-10) appears as a sinusoidal function of ϕ_e (the displacement from the q axis) with its maximum value occurring at $\phi_e = \theta_{esi}(0)$ which is the phase angle of the balanced steady-state stator currents. In other words, we are observing one half of Tesla's rotating magnetic field as we run at ω_e in the counterclockwise direction around the air gap. Since ω_e is the electrical angular velocity of I_{as} and I_{bs}, this is referred to as the synchronously rotating frame of reference. The view from the synchronous rotating frame is depicted in Fig. 2.2-3 where $\theta_{esi}(0)$ is negative for inductive circuits (see Example 1.A). Applying the rule given in Chapter 1, a positive mmf_s^e is a south pole due to the stator

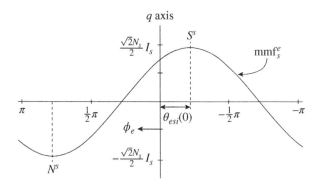

Figure 2.2-3 Tesla's rotating magnetic field $\left(\text{mmf}_s^e\right)$ viewed from $-\pi$ to π by an observer rotating counterclockwise about the air gap at ω_e with $\theta(0) = 0$ and $\theta_{esi}(0)$ is negative.

currents (S^s) with the maximum intensity at $\phi_e = \theta_{esi}(0)$ which is the phase angle of the currents I_{as} and I_{bs}. Flux entering the member with the winding. A negative mmf_s^e represents a stator north pole (N^s). Flux leaving the member with the winding.

We have observed Tesla's rotating magnetic field from the synchronous reference frame which is a frame of reference rotating at ω_e. We will find that the synchronous reference frame allows us to work with dc variables and eliminates position-dependent inductances. This simplifies the analysis and enables a clear visualization of the operation of ac electric machines.

This is not the first time we have viewed a sinusoidal variation from a reference frame rotating at ω_e. In Section 1.2, Example 1.A, it was illustrated that a phasor rotated at ω_e in the counterclockwise direction and when we consider the phasor as a constant, we have either stopped the rotation or we are running counterclockwise at ω_e. The position of the winding and the phasor of the current locate the poles. Therefore, we can superimpose the poles on the phasor diagram. We will talk more about poles and the phasor diagram in later sections.

It is important not to confuse the magnetic axes of the stator windings (as and bs axes) which are stationary, with the definition of the phasors representing the stator currents ($\widetilde{I}_{as}$ and $\widetilde{I}_{bs}$). However, it is interesting to note the relative position of the as axis and bs axis versus the relative position of $\widetilde{I}_{as}$ and $\widetilde{I}_{bs}$ on a phasor diagram for a constant, counterclockwise rotating mmf_s to occur. The bs axis is displaced $\pi/2$ ahead of the as axis, as illustrated in Fig. 2.2-1; however, from (2.2-3) and (2.2-4), we see that $\widetilde{I}_{bs}$ lags $\widetilde{I}_{as}$ by $\pi/2$. The direction of rotation of the air-gap mmf (mmf_s) may readily be determined by forgetting about $\theta_{esi}(0)$ and locate the position of $\frac{N_s}{2}\sqrt{2}I_s$ when time has progressed to where I_{bs} is maximum and I_{as} is zero ($\omega_e t = \pi/2$). That is, the maximum mmf_s has rotated counterclockwise from the *as* axis when I_{as} is maximum to the *bs* axis when I_{bs} is maximum. Note, that the negative of both (2.2-3) and (2.2-4) would also produce a counterclockwise mmf$_s$. In this case, $\tilde{I}_{as}$ would be at 180° and $\tilde{I}_{bs}$ at 90°. In other words, maximum I_{as} followed $\pi/2$ electrical degrees by negative maximum I_{bs} produces counterclockwise rotation of mmf$_s$ in the device shown in Fig. 2.2-1.

Example 2.A   Components of mmf$_s^e$

Express mmf$_{as}^e$ and mmf$_{bs}^e$ with $\theta(0) = 0$. Add the resulting relationships to obtain mmf$_s^e$ which is (2.2-10). The currents may be expressed as

$$I_{as} = \sqrt{2}I_s \cos[\omega_e t + \theta_{esi}(0)] \tag{2A-1}$$

$$I_{bs} = \sqrt{2}I_s \sin[\omega_e t + \theta_{esi}(0)] \tag{2A-2}$$

To view the mmf as an observer running at ω_e counterclockwise with the q axis (Fig. 2.2-2),

$$\phi_s = \omega_e t + \phi_e \tag{2A-3}$$

where ϕ_e is the displacement from the q axis when it is rotating at ω_e with $\theta(0) = 0$ ($\theta_e = \omega_e t$). Thus, from (2.2-7),

$$
\begin{aligned}
\text{mmf}_{as}^e &= \frac{N_s}{2}\sqrt{2}I_s\{\cos[\omega_e t + \theta_{esi}(0)]\cos(\omega_e t + \phi_e)\} \\
&= \frac{N_s}{2}\sqrt{2}I_s\left(\frac{1}{2}\right)\{\cos[2\omega_e t + \theta_{esi}(0) + \phi_e] + \cos[\theta_{esi}(0) - \phi_e]\}
\end{aligned}
\tag{2A-4}
$$

From (2.2-1), we can express mmf$_{bs}^e$ as

$$\text{mmf}_{bs}^e = \frac{N_s}{2}\sqrt{2}I_s\left(\frac{1}{2}\right)\{\cos[\theta_{esi}(0) - \phi_e] - \cos[2\omega_e t + \theta_{esi}(0) + \phi_e]\} \tag{2A-5}$$

Now, from (2.2-2),

$$
\begin{aligned}
\text{mmf}_s^e &= \text{mmf}_{as}^e + \text{mmf}_{bs}^e \\
&= \frac{N_s}{2}\sqrt{2}I_s \cos[\theta_{esi}(0) - \phi_e]
\end{aligned}
\tag{2A-6}
$$

The double frequency terms cancel and we obtain (2A-6) which is (2.2-10) and it is a constant if $\theta_{esi}(0)$ and ϕ_e are constants.

2.2.2 Two-Pole Three-Phase Stator

The arrangement of the stator windings of a two-pole three-phase device is shown in Fig. 2.2-4. It is noted that the three-phase machine requires only three wires while the two-phase requires four wires. Moreover, the three-phase systems can transmit 150% more power than a two-phase system with one less wire.

The windings are connected in wye and they are identical, sinusoidally distributed with N_s equivalent turns and with their magnetic axes displaced by $\frac{2}{3}\pi$. The positive directions of the magnetic axes are selected so as to achieve counterclockwise rotation of the rotating air-gap mmf with balanced stator currents of the *abc* sequence. We shall see this in just a moment. The air-gap mmfs established by the stator phases may be expressed by inspection of Fig. 2.2-4. In particular,

$$\text{mmf}_{as} = \frac{N_s}{2} i_{as} \cos \phi_s \tag{2.2-11}$$

$$\text{mmf}_{bs} = \frac{N_s}{2} i_{bs} \cos \left(\phi_s - \frac{2}{3}\pi \right) \tag{2.2-12}$$

$$\text{mmf}_{cs} = \frac{N_s}{2} i_{cs} \cos \left(\phi_s + \frac{2}{3}\pi \right) \tag{2.2-13}$$

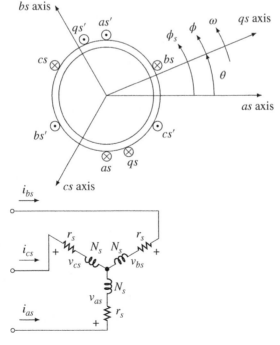

Figure 2.2-4 Elementary two-pole three-phase sinusoidally distributed stator windings. Outer housing is not shown.

As before, N_s is the number of turns of the equivalent sinusoidally distributed stator windings and ϕ_s is the angular displacement about the stator measured from the *as* axis. The air-gap mmf_s is

$$\text{mmf}_s = \frac{N_s}{2}\left[i_{as}\cos\phi_s + i_{bs}\cos\left(\phi_s - \frac{2}{3}\pi\right) + i_{cs}\cos\left(\phi_s + \frac{2}{3}\pi\right)\right] \quad (2.2\text{-}14)$$

For balanced steady-state conditions, the stator currents for an *abc* sequence may be expressed as

$$I_{as} = \sqrt{2}I_s\cos\left[\omega_e t + \theta_{esi}(0)\right] \quad (2.2\text{-}15)$$

$$I_{bs} = \sqrt{2}I_s\cos\left[\omega_e t - \frac{2}{3}\pi + \theta_{esi}(0)\right] \quad (2.2\text{-}16)$$

$$I_{cs} = \sqrt{2}I_s\cos\left[\omega_e t + \frac{2}{3}\pi + \theta_{esi}(0)\right] \quad (2.2\text{-}17)$$

Substituting (2.2-15) through (2.2-17) into (2.2-14) and using (2.2-6) for ϕ_s and then after some trigonometric manipulations, we can obtain an expression for Tesla's rotating magnetic field established by balanced steady-state stator currents

$$\text{mmf}_s = \frac{N_s}{2}\sqrt{2}I_s\frac{3}{2}\cos\left[(\omega_e - \omega)t + \theta_{esi}(0) - \theta(0) - \phi\right], \quad (2.2\text{-}18)$$

where ω, $\theta(0)$, and ϕ are defined as they were for two-phase machine and are shown in Fig. 2.2-4. If mmf_s for the three-phase device given by (2.2-18) is compared with mmf_s for a two-phase device given by (2.2-8), we see that they are identical except that the amplitude for the three-phase device is $\frac{3}{2}$ times that of a two-phase device. We need not repeat the work we did leading up to Fig. 2.2-3. The only difference would be the 3/2 factor. Therefore, it is reasonable to treat the two-phase machine first since the rotating magnetic fields differ only by a constant.

It is important to note that, with the assumed positive directions of the magnetic axes, a counterclockwise rotating air-gap mmf is obtained with a three-phase set of balanced stator currents of the *abc* sequence. As in the two-phase case, it is also important to note the relative positions of the magnetic axes versus the relative positions of the phasors representing the currents in order to establish a constant amplitude counterclockwise rotating air-gap mmf. From (2.2-11) through (2.2-13) or Fig. 2.2-4, we see that the *bs* axis is stationary at 120°, whereas the *cs* axis is stationary at −120°. From (2.2-15) through (2.2-17), $\tilde{I}_{as}$, $\tilde{I}_{bs}$, and $\tilde{I}_{cs}$ are 120° out of phase; however, in order for the constant-amplitude air-gap mmf to rotate in the counterclockwise direction, $\tilde{I}_{bs}$ lags $\tilde{I}_{as}$ by 120° and $\tilde{I}_{cs}$ lags $\tilde{I}_{as}$ by 240°. In other words, maximum positive I_{as} is followed by maximum positive I_{bs} and then maximum positive I_{cs} for an *abc* sequence. With the magnetic axis positioned as shown in Fig. 2.2-4, this sequence of maximum currents produces a counterclockwise rotation of the air-gap mmf.

SP2.2-1. We have expressed mmf_s^s and mmf_s^e. Express mmf_s^r; a reference frame rotating counterclockwise at the angular velocity of the rotor, ω_r. [(2.2-8) for a two-phase stator or (2.2-18) for three-phase with $\omega = \omega_r$ and ϕ replaced by ϕ_r]

SP2.2-2. Express mmf_s^s for a *ba* sequence of a two-phase stator.

2.3 Reference Frame Theory

R. H. Park set forth a change of variables in 1929 that eliminated the position-varying inductances of a synchronous machine [2]. His work, which was for sinusoidal distributed windings and generator action with positive current out of the stator convenient for supplying loads connected to the power grid. Nevertheless, it revolutionized machine analysis. This was the beginning of *Reference Frame Theory* and modern electric machine analysis; however, it was not appreciated until the advent of the computer and later power electronics and electric drives. It is now a necessity to design and analyze drive systems. The use of reference frame theory to analyze electric machines is the main thrust of this text; no longer can we focus only on steady-state operation of electric machines and drives, the student must become knowledgeable of reference frame theory in the early courses. It is no longer an option, it has become a necessity.

Only recently was the connection between Tesla and Park discovered [3]. As it turns out, reference frame theory is nothing more than establishing the variables associated with the fictitious circuits that produce Tesla's balanced rotating magnetic field as viewed from that reference frame.

2.3.1 Two-Phase Transformation

The transformation or change of variables can be readily established from (2.2-2), (2.2-6), and (2.2-7). These equations are repeated here to start our derivation.

$$\text{mmf}_s = \frac{N_s}{2}(i_{as}\cos\phi_s + i_{bs}\sin\phi_s) \tag{2.3-1}$$

$$\phi_s = \theta + \phi \tag{2.3-2}$$

Substituting (2.3-2) into (2.3-1) yields

$$\text{mmf}_s = \frac{N_s}{2}[i_{as}\cos(\theta + \phi) + i_{bs}\sin(\theta + \phi)] \tag{2.3-3}$$

Equation (2.3-3) may be written, using basic trigonometry, as

$$\text{mmf}_s = \frac{N_s}{2} \cos\phi (i_{as} \cos\theta + i_{bs} \sin\theta) + \frac{N_s}{2} \sin\phi (-i_{as} \sin\theta + i_{bs} \cos\theta) \quad (2.3\text{-}4)$$

As shown in Fig. 2.3-1 and from (2.3-4), $N_s \cos\phi$ and $N_s \sin\phi$ result from two orthogonal sinusoidally distributed windings that are fictitious except when $\theta = 0$, the stationary reference frame.

The current flowing in the winding hovered around the q axis ($\phi = 0$) is the arbitrary reference frame variable i_{qs} which is expressed as

$$i_{qs} = i_{as} \cos\theta + i_{bs} \sin\theta \quad (2.3\text{-}5)$$

The second winding axis can be at $\phi = \pm\frac{\pi}{2}$. We will let $\phi = -\frac{\pi}{2}$, then the second arbitrary reference frame current which we will call i_{ds}. It is expressed as

$$i_{ds} = i_{as} \sin\theta - i_{bs} \cos\theta \quad (2.3\text{-}6)$$

We can now write (2.3-4) in the arbitrary reference frame variables as

$$\text{mmf}_s = \frac{N_s}{2} i_{qs} \cos\phi - \frac{N_s}{2} i_{ds} \sin\phi \quad (2.3\text{-}7)$$

The change of variables or transformation to the arbitrary reference frame expressed in matrix form is

$$\begin{bmatrix} i_{qs} \\ i_{ds} \end{bmatrix} = \begin{bmatrix} \cos\theta & \sin\theta \\ \sin\theta & -\cos\theta \end{bmatrix} \begin{bmatrix} i_{as} \\ i_{bs} \end{bmatrix} \quad (2.3\text{-}8)$$

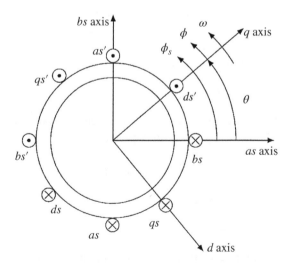

Figure 2.3-1 Elementary two-pole two-phase stator with a two-phase q and d axes.

This same transformation also applies to voltage and flux linkages, thus

$$\begin{bmatrix} f_{qs} \\ f_{ds} \end{bmatrix} = \begin{bmatrix} \cos\theta & \sin\theta \\ \sin\theta & -\cos\theta \end{bmatrix} \begin{bmatrix} f_{as} \\ f_{bs} \end{bmatrix} \tag{2.3-9}$$

where f can be voltage, current, or flux linkage. We can also write (2.3-9) as

$$\mathbf{f}_{qds} = \mathbf{K}_s \mathbf{f}_{abs} \tag{2.3-10}$$

Now,

$$\frac{d\theta}{dt} = \omega(t) \tag{2.3-11}$$

where ω is the unspecified speed of the arbitrary reference frame; we are free to assign it any value. Also, $\mathbf{K}_s$ is considered as the transformation to the arbitrary reference frame and with $\phi = -\frac{\pi}{2}$, $(\mathbf{K})^{-1} = \mathbf{K}_s$. In this text, we will only concern ourselves with three specific reference frames: stationary, $\omega = 0$; the rotor, $\omega = \omega_r$; and the synchronously rotating, $\omega = \omega_e$. It is interesting that for reference frames rotating at $\omega < \omega_e$, the mmf$_s$ rotates counterclockwise and for reference frames rotating at $\omega > \omega_e$, mmf$_s$ rotates clockwise. In other words, mmf$_s$ rotates toward synchronous speed [4].

2.3.2 Three-Phase Transformation

For the three-phase transformation, we will start with (2.2-14) which is

$$\text{mmf}_s = \frac{N_s}{2}\left[i_{as}\cos\phi_s + i_{bs}\cos\left(\phi_s - \frac{2}{3}\pi\right) + i_{cs}\cos\left(\phi_s + \frac{2}{3}\pi\right) \right] \tag{2.3-12}$$

Substituting (2.3-2) for ϕ_s yields

$$\begin{aligned} \text{mmf}_s &= \frac{N_s}{2}\cos\phi\left[i_{as}\cos\theta + i_{bs}\cos\left(\theta - \frac{2}{3}\pi\right) + i_{cs}\cos\left(\theta + \frac{2}{3}\pi\right) \right] \\ &- \frac{N_s}{2}\sin\phi\left[i_{as}\sin\theta + i_{bs}\sin\left(\theta - \frac{2}{3}\pi\right) + i_{cs}\sin\left(\theta + \frac{2}{3}\pi\right) \right] \end{aligned} \tag{2.3-13}$$

Here, i_{qs} is the [ ] multiplying $\cos\phi$ with $\phi = 0$ and i_{ds} is the [ ] multiplying $\sin\phi$ with $\phi = -\frac{\pi}{2}$. In addition to the q and d variables, a third variable, the zero variable, is introduced as

$$f_{0s} = \frac{1}{3}(f_{as} + f_{bs} + f_{cs}) \tag{2.3-14}$$

The transformation to the arbitrary reference frame becomes

$$\mathbf{f}_{qd0s} = \mathbf{K}_s \mathbf{f}_{abcs} \tag{2.3-15}$$

where in the case of the three-phase machine

$$\left(\mathbf{f}_{qd0s}\right)^T = \begin{bmatrix} f_{qs} & f_{ds} & f_{0s} \end{bmatrix} \tag{2.3-16}$$

$$\left(\mathbf{f}_{abcs}\right)^T = \begin{bmatrix} f_{as} & f_{bs} & f_{cs} \end{bmatrix} \tag{2.3-17}$$

$$\mathbf{K}_s = \frac{2}{3} \begin{bmatrix} \cos\theta & \cos\left(\theta - \frac{2}{3}\pi\right) & \cos\left(\theta + \frac{2}{3}\pi\right) \\ \sin\theta & \sin\left(\theta - \frac{2}{3}\pi\right) & \sin\left(\theta + \frac{2}{3}\pi\right) \\ \frac{1}{2} & \frac{1}{2} & \frac{1}{2} \end{bmatrix} \tag{2.3-18}$$

$$\left(\mathbf{K}_s\right)^{-1} = \begin{bmatrix} \cos\theta & \sin\theta & 1 \\ \cos\left(\theta - \frac{2}{3}\pi\right) & \sin\left(\theta - \frac{2}{3}\pi\right) & 1 \\ \cos\left(\theta + \frac{2}{3}\pi\right) & \sin\left(\theta + \frac{2}{3}\pi\right) & 1 \end{bmatrix} \tag{2.3-19}$$

where

$$\frac{d\theta}{dt} = \omega(t) \tag{2.3-20}$$

Note that the f_{0s} variable does not contain θ and is zero for balanced conditions. The 2/3 factor in (2.3-18) was introduced by Park. It is not suggested by Tesla. Due to this 2/3 factor in Park's transformation, which makes the magnitude of the variables equal to those of the two-phase system, we must multiply the power and torque by 3/2 in order to correct this reduction in the power calculated using Park's variables. In other words, Park's transformation makes a three-phase machine a two-phase machine. We will take care of all this when we consider the individual machines.

Example 2.B   Applied Voltages in Different Reference Frames

We will assume that

$$v_{as} = \sqrt{2}V_s \cos\theta_{esv} \tag{2B-1}$$

$$v_{bs} = \sqrt{2}V_s \sin\theta_{esv} \tag{2B-2}$$

$$\theta_{esv} = \omega_e t + \theta_{esv}(0) \tag{2B-3}$$

Throughout this text the applied voltages will be of the form given by (2B-1) through (2B-3). Although V_s and θ_{esv} are generally constant. It is desirable to express the applied voltages, v_{qs} and v_{ds} in the arbitrary, stationary, and synchronously rotating reference frames. For this purpose, let

$$\theta = \omega t + \theta(0) \tag{2B-4}$$

Substituting (2B-1) through (2B-4) into (2.3-9) and after some work, we have the expression for the applied voltages in the arbitrary reference frame

$$v_{qs} = \sqrt{2}V_s \cos\left[(\omega_e - \omega)t + \theta_{esv}(0) - \theta(0)\right] \tag{2B-5}$$

$$v_{ds} = -\sqrt{2}V_s \sin\left[(\omega_e - \omega)t + \theta_{esv}(0) - \theta(0)\right] \tag{2B-6}$$

where $\theta_{esv}(0)$ is the phase angle of v_{as} and v_{bs} and $\theta(0)$ is zero unless otherwise specified.

For the stationary reference frame $\omega = 0$ and

$$v_{qs}^s = v_{as} \tag{2B-7}$$

$$v_{ds}^s = -v_{bs} \tag{2B-8}$$

where the superscript s denotes variables in the stationary reference frame. In the synchronously rotating reference frame, $\omega = \omega_e$,

$$v_{qs}^e = \sqrt{2}V_s \cos\theta_{esv}(0) \tag{2B-9}$$

$$v_{ds}^e = -\sqrt{2}V_s \sin\theta_{esv}(0) \tag{2B-10}$$

where the raised e denotes variables in the synchronously rotating reference frame. It is important to note that, in the steady state, v_{qs}^e and v_{ds}^e are constant dc, applied voltages. It is a good bet that the currents i_{qs}^e and i_{ds}^e will also be dc in the steady state.

SP2.3-1. Express v_{qs} and v_{ds} in the arbitrary reference frame for a balanced three-phase system. The trigonometric identities in Appendix A will help. [(2B-5) and (2B-6)]

SP2.3-2. Assume v_{ds} and i_{ds} are both zero. Show that we must multiply power calculated using Parks transformation by 3/2. [$P_{3\phi} = (3/2)P_{2\phi} = (3/2)2V_sI_s = 3V_sI_s$]

SP2.3-3. Show that for balanced conditions f_{0s} is zero. [$f_{as} + f_{bs} + f_{cs} = 0$]

SP2.3-4. Use (2B-5) and (2B-6) to express v_{qs}^r and v_{ds}^r. [Let $\omega = \omega_r$]

SP2.3-5. Why is the transformation used here convenient for drive systems? [drive machines are motors]

2.4 Stator Voltage and Flux Linkage Equations in the Arbitrary Reference Frame and the Instantaneous Phasor

In this section, we will transform the voltages, currents, and flux linkage equations for the two- and three-phase stator to the arbitrary reference frame. The results will be used in later chapters to investigate the behavior of induction and synchronous machines.

2.4.1 Two-Phase Stator

Let us consider Fig. 2.3-1, where current is positive in the direction of voltage drop, motor action

$$v_{as} = r_s i_{as} + \frac{d\lambda_{as}}{dt} \tag{2.4-1}$$

$$v_{bs} = r_s i_{bs} + \frac{d\lambda_{bs}}{dt} \tag{2.4-2}$$

which may be written as

$$\mathbf{v}_{abs} = \mathbf{r}_s \mathbf{i}_{abs} + p\lambda_{abs} \tag{2.4-3}$$

where the first term on the right-hand side comes from Ohm's law and the second term from Faraday's law, Also, p is the operator d/dt. From (2.3-10),

$$\mathbf{f}_{abs} = (\mathbf{K}_s)^{-1} \mathbf{f}_{qds} \tag{2.4-4}$$

where $(\mathbf{K})^{-1} = \mathbf{K}_s$ for the d axis behind the q axis by $\frac{\pi}{2}$ radians or $\phi = -\frac{\pi}{2}$ in (2.3-4). Substituting into (2.4-3) yields

$$(\mathbf{K}_s)^{-1} \mathbf{v}_{qds} = \mathbf{r}_s (\mathbf{K}_s)^{-1} \mathbf{i}_{qds} + p\left[(\mathbf{K}_s)^{-1} \lambda_{qds}\right] \tag{2.4-5}$$

Multiplying each side of (2.4-5) by $\mathbf{K}_s$ gives

$$\mathbf{v}_{qds} = \mathbf{K}_s \mathbf{r}_s (\mathbf{K}_s)^{-1} \mathbf{i}_{qds} + \mathbf{K}_s p\left[(\mathbf{K}_s)^{-1}\right] \lambda_{qds} + \mathbf{K}_s (\mathbf{K}_s)^{-1} p\lambda_{qds} \tag{2.4-6}$$

where the product rule has been used to express the second term on the right-hand side of (2.4-5).

Now, term by term of (2.4-6)

$$\begin{aligned} \mathbf{K}_s \mathbf{r}_s (\mathbf{K}_s)^{-1} \mathbf{i}_{qds} &= \mathbf{K}_s r_s \mathbf{I} (\mathbf{K}_s)^{-1} \mathbf{i}_{qds} \\ &= r_s \mathbf{K}_s (\mathbf{K}_s)^{-1} \mathbf{i}_{qds} = \mathbf{r}_s \mathbf{i}_{qds} \end{aligned} \tag{2.4-7}$$

$$\mathbf{K}_s p(\mathbf{K}_s)^{-1}\lambda_{qds} = \mathbf{K}_s\omega \begin{bmatrix} -\sin\theta & \cos\theta \\ \cos\theta & \sin\theta \end{bmatrix} \begin{bmatrix} \lambda_{qs} \\ \lambda_{ds} \end{bmatrix}$$

$$= \omega \begin{bmatrix} 0 & 1 \\ -1 & 0 \end{bmatrix} \begin{bmatrix} \lambda_{qs} \\ \lambda_{ds} \end{bmatrix} \tag{2.4-8}$$

$$= \omega \begin{bmatrix} \lambda_{ds} \\ -\lambda_{qs} \end{bmatrix} = \omega\lambda_{dqs}$$

$$\mathbf{K}_s(\mathbf{K_s})^{-1}p\lambda_{qds} = p\lambda_{qds} \tag{2.4-9}$$

The voltage equation in the arbitrary reference frame becomes

$$\mathbf{v}_{qds} = \mathbf{r}_s\mathbf{i}_{qds} + \omega\lambda_{dqs} + p\lambda_{qds} \tag{2.4-10}$$

where $\omega\lambda_{dqs}$ is (2.4-8).

In expanded form

$$v_{qs} = r_s i_{qs} + \omega\lambda_{ds} + p\lambda_{qs} \tag{2.4-11}$$

$$v_{ds} = r_s i_{ds} - \omega\lambda_{qs} + p\lambda_{ds} \tag{2.4-12}$$

These are the voltage equations associated with the fictitious windings in the arbitrary reference frame. For example, if we assign ($\omega = 0$), we yield the voltage equations in the stationary reference frame

$$v_{qs}^s = r_s i_{qs}^s + p\lambda_{qs}^s \tag{2.4-13}$$

$$v_{ds}^s = r_s i_{ds}^s + p\lambda_{ds}^s \tag{2.4-14}$$

where the raised s denotes variables in the stationary or stator reference frame. In the rotor reference frame, ($\omega = \omega_r$), and

$$v_{qs}^r = r_s i_{qs}^r + \omega_r\lambda_{ds}^r + p\lambda_{qs}^r \tag{2.4-15}$$

$$v_{ds}^r = r_s i_{ds}^r - \omega_r\lambda_{qs}^r + p\lambda_{ds}^r \tag{2.4-16}$$

where the raised r denotes variables in the rotor reference frame. In the synchronously rotating reference frame, ($\omega = \omega_e$),

$$v_{qs}^e = r_s i_{qs}^e + \omega_e\lambda_{ds}^e + p\lambda_{qs}^e \tag{2.4-17}$$

$$v_{ds}^e = r_s i_{ds}^e - \omega_e\lambda_{qs}^e + p\lambda_{ds}^e \tag{2.4-18}$$

where the raised e denotes variables in the synchronously ω_e rotating reference frame. These equations provide the rotating magnetic field as viewed from the stationary, rotor, and synchronous reference frames. The applied voltages in the arbitrary reference frame are given for balanced conditions by (2B-5) and (2B-6). We have not defined λ_{qs} and λ_{ds}, therefore the above voltage equations are valid for linear or nonlinear magnetic systems.

Example 2.C   Voltage Equations with d-axis located at $\phi = \dfrac{\pi}{2}$

The transformation, $\mathbf{K}_s$, for $\phi = \dfrac{\pi}{2}$ in (2.3-4) is

$$
\begin{bmatrix} f_{qs} \\ f_{ds} \end{bmatrix} = \begin{bmatrix} \cos\theta & \sin\theta \\ -\sin\theta & \cos\theta \end{bmatrix} \begin{bmatrix} f_{as} \\ f_{bs} \end{bmatrix}
$$

$$
= \mathbf{K}_s \begin{bmatrix} f_{as} \\ f_{bs} \end{bmatrix}
$$

(2C-1)

In this case,

$$
(\mathbf{K}_s)^{-1} = \begin{bmatrix} \cos\theta & \sin\theta \\ \sin\theta & -\cos\theta \end{bmatrix}
$$

(2C-2)

Thus, $(\mathbf{K}_s)^{-1} \neq \mathbf{K}_s$. Now, (2.4-7) and (2.4-9) do not change, that is,

$$
\mathbf{K}_s \mathbf{r}_s (\mathbf{K}_s)^{-1} \mathbf{i}_{qds} = \mathbf{r}_s \mathbf{i}_{qds}
$$

(2C-3)

$$
\mathbf{K}_s (\mathbf{K}_s)^{-1} p\boldsymbol{\lambda}_{qds} = p\boldsymbol{\lambda}_{qds}
$$

(2C-4)

The speed voltage changes

$$
\begin{aligned}
\mathbf{K}_s p (\mathbf{K}_s)^{-1} \boldsymbol{\lambda}_{qds} &= \mathbf{K}_s \omega \begin{bmatrix} -\sin\theta & -\cos\theta \\ \cos\theta & -\sin\theta \end{bmatrix} \begin{bmatrix} \lambda_{qs} \\ \lambda_{ds} \end{bmatrix} \\
&= \omega \begin{bmatrix} \cos\theta & \sin\theta \\ -\sin\theta & \cos\theta \end{bmatrix} \begin{bmatrix} -\sin\theta & -\cos\theta \\ \cos\theta & -\sin\theta \end{bmatrix} \begin{bmatrix} \lambda_{qs} \\ \lambda_{ds} \end{bmatrix} \\
&= \omega \begin{bmatrix} 0 & -1 \\ 1 & 0 \end{bmatrix} \begin{bmatrix} \lambda_{qs} \\ \lambda_{ds} \end{bmatrix} \\
&= \omega \begin{bmatrix} -\lambda_{ds} \\ \lambda_{qs} \end{bmatrix}
\end{aligned}
$$

(2C-5)

The voltage equations in the arbitrary reference frame become

$$
v_{qs} = r_s i_{qs} - \omega\lambda_{ds} + p\lambda_{qs}
$$

(2C-6)

$$
v_{ds} = r_s i_{ds} + \omega\lambda_{qs} + p\lambda_{ds}
$$

(2C-7)

The sign associated with the speed voltages is reversed.

The flux linkage equations for a magnetically linear two-phase stator are given by (1.5-12) through (1.5-15). From (1.5-13),

$$
(\mathbf{K}_s)^{-1} \boldsymbol{\lambda}_{qds} = \mathbf{L}_s (\mathbf{K}_s)^{-1} \mathbf{i}_{qds}
$$

(2.4-19)

or

$$
\begin{aligned}
\lambda_{qds} &= \mathbf{K}_s \mathbf{L}_s (\mathbf{K}_s)^{-1} \mathbf{i}_{qds} \\
&= \mathbf{K}_s L_{ss} \mathbf{I} (\mathbf{K}_s)^{-1} \mathbf{i}_{qds} \\
&= L_{ss} \mathbf{K}_s (\mathbf{K}_s)^{-1} \mathbf{i}_{qds} \\
&= \mathbf{L}_s \mathbf{i}_{qds}
\end{aligned}
\tag{2.4-20}
$$

2.4.2 Three-Phase Stator

The three-phase wye-connected stator is given in Fig. 1.5-2. The voltage equation is given by (1.5-25) through (1.5-28). Transforming to the arbitrary reference frame, (1.5-28) becomes

$$
\begin{aligned}
(\mathbf{K}_s)^{-1} \mathbf{v}_{qd0s} = {}& \mathbf{r}_s (\mathbf{K}_s)^{-1} \mathbf{i}_{qd0s} + p \left[(\mathbf{K})^{-1} \right] \lambda_{qd0s} \\
& + (\mathbf{K}_s)^{-1} p \lambda_{qd0s}
\end{aligned}
\tag{2.4-21}
$$

where $\mathbf{K}_s$ and $(\mathbf{K}_s)^{-1}$ are given by (2.3-18) and (2.3-19), respectively. Following the work for the two-phase stator, (2.4-10) becomes

$$
\mathbf{v}_{qd0s} = \mathbf{r}_s \mathbf{i}_{qd0s} + \omega \lambda_{dq0s} + p \lambda_{qd0s}
\tag{2.4-22}
$$

In expanded form

$$
v_{qs} = r_s i_{qs} + \omega \lambda_{ds} + p \lambda_{qs}
\tag{2.4-23}
$$

$$
v_{ds} = r_s i_{ds} - \omega \lambda_{qs} + p \lambda_{ds}
\tag{2.4-24}
$$

$$
v_{0s} = r_s i_{0s} + p \lambda_{0s}
\tag{2.4-25}
$$

Except for v_{0s} these equations are of identical form to the two-phase stator. Moreover, v_{0s} is zero for balanced conditions and for a wye-connected symmetrical stator.

The flux linkage equations for a magnetically linear system become

$$
(\mathbf{K})^{-1} \lambda_{qd0s} = \mathbf{L}_s (\mathbf{K}_s)^{-1} \mathbf{i}_{qd0s}
\tag{2.4-26}
$$

where $\mathbf{K}_s$ is (2.3-18), $(\mathbf{K}_s)^{-1}$ is (2.3-19), and $\mathbf{L}_s$ is (1.5-30), We can write (2.4-26) as

$$
\lambda_{qd0s} = \mathbf{K}_s \mathbf{L}_s (\mathbf{K}_s)^{-1} \mathbf{i}_{qd0s}
\tag{2.4-27}
$$

Now, for a wye-connection or for balanced conditions

$$
i_{as} + i_{bs} + i_{cs} = 0
\tag{2.4-28}
$$

and if we let $L_{Ms} = \dfrac{3}{2} L_{ms}$, we can write $\mathbf{L}_s$ as

$$\mathbf{L}_s = \begin{bmatrix} L_{ls} + L_{Ms} & 0 & 0 \\ 0 & L_{ls} + L_{Ms} & 0 \\ 0 & 0 & L_{ls} + L_{Ms} \end{bmatrix} = (L_{ls} + L_{Ms})\,\mathbf{I} \qquad (2.4\text{-}29)$$

Thus,

$$\begin{aligned} \lambda_{qd0s} &= \mathbf{K}_s(L_{ls} + L_{Ms})\,\mathbf{I}(\mathbf{K}_s)^{-1}\mathbf{i}_{qd0s} \\ &= (L_{ls} + L_{Ms})\,\mathbf{K}_s\mathbf{I}(\mathbf{K}_s)^{-1}\mathbf{i}_{qd0s} \end{aligned} \qquad (2.4\text{-}30)$$

For wye-connected stator windings where $i_{as} + i_{bs} + i_{cs} = 0$, this becomes

$$\begin{bmatrix} \lambda_{qs} \\ \lambda_{ds} \\ \lambda_{0s} \end{bmatrix} = \begin{bmatrix} L_{ls} + L_{Ms} & 0 & 0 \\ 0 & L_{ls} + L_{Ms} & 0 \\ 0 & 0 & L_{ls} \end{bmatrix} \begin{bmatrix} i_{qs} \\ i_{ds} \\ i_{0s} \end{bmatrix} \qquad (2.4\text{-}31)$$

The 3x3 term is $L_{ls} + L_{Ms}$ in (2.4-29); however, the $L_{Ms}i_{0s}$ term is coupled between the *as*, *bs*, and *cs* axes which sums to zero. The leakage inductances, L_{ls}, are not coupled.

2.4.3 Instantaneous and Steady-State Phasors

The synchronously rotating reference frame can be thought of as a synchronously rotating complex plane as shown in Fig. 2.4-1, whereupon we can think of f_{qs}^e and f_{ds}^e as $\tilde{f}_{as}$, an instantaneous phasor of phase *as* variables. This would include the steady-state and transient response for balanced conditions for a two- or three-phase stator. From Fig. 2.4-1, we can write

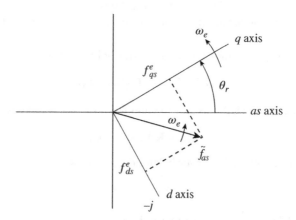

Figure 2.4-1 The *q* and *d* complex plane.

$$\tilde{f}_{as}(t) = f^e_{qs}(t) - jf^e_{ds}(t) \tag{2.4-32}$$

Substituting (2.4-17) and (2.4-18) into (2.4-32), whereupon for both the two- or three-phase stator

$$v^e_{qs} - jv^e_{ds} = r_s i^e_{qs} + \omega_e \lambda^e_{ds} + p\lambda^e_{qs} - j\left(r_s i^e_{ds} - \omega_e \lambda^e_{qs} + p\lambda^e_{ds}\right) \tag{2.4-33}$$

In terms of instantaneous phasors

$$\tilde{v}_{as} = r_s \tilde{i}_{as} + j\omega_e \tilde{\lambda}_{as} + p\tilde{\lambda}_{as} \tag{2.4-34}$$

Once the transient subsides, the i^e_{qs} and i^e_{ds} become constant, $p\tilde{\lambda}_{as}$ becomes zero, and (2.4-34) becomes

$$\tilde{v}_{as} = r_s \tilde{i}_{as} + j\omega_e \tilde{\lambda}_{as} \tag{2.4-35}$$

Now, (2.4-35) is expressed in instantaneous variables and must be divided by $\sqrt{2}$ to be expressed in terms of steady-state phasors. Also, for a magnetically linear two-phase stator and the *rL* circuit being considered

$$\tilde{\lambda}_{as} = L_{ss} \tilde{i}_{as} \tag{2.4-36}$$

thus, (2.4-35) becomes

$$\tilde{V}_{as} = (r + j\omega_e L_{ss})\tilde{I}_{as} \tag{2.4-37}$$

for a two-phase stator $L_{ss} = L_{ls} + L_{ms}$, for a three-phase L_{ms} is replaced by L_{Ms}. Now,

$$\sqrt{2}\tilde{F}_{as} = F^e_{qs} - jF^e_{ds} \tag{2.4-38}$$

where the uppercase letters denote steady-state variables.

For a two-phase *ab* sequence,

$$\sqrt{2}\tilde{F}_{bs} = -\sqrt{2}j\tilde{F}_{as} \\ = -F^e_{ds} - jF^e_{qs} \tag{2.4-39}$$

For a three-phase *abc* sequence, $\tilde{F}_{as}$ is (2.4-38) and

$$\tilde{F}_{bs} = \tilde{F}_{as}\underline{/-120°} \tag{2.4-40}$$

$$\tilde{F}_{cs} = \tilde{F}_{as}\underline{/+120°} \tag{2.4-41}$$

We will make extensive use of the instantaneous phasor when dealing with the dynamic response of machine performance.

Example 2.D   The Instantaneous and Steady-State Phasors

Let $r_s = 3.4\,\Omega$, $L_{ls} = 1.1\,\text{mH}$, and $L_{ms} = 11\,\text{mH}$. Assume $\omega_e = 377\,\text{rad/s}$ and for a two-phase stator

$$\lambda_{qs}^e = L_{ss}i_{qs}^e \tag{2D-1}$$

$$\lambda_{ds}^e = L_{ss}i_{ds}^e \tag{2D-2}$$

Express $\widetilde{v}_{as}$ and $\widetilde{V}_{as}$. From (2.4-34),

$$\begin{aligned}
\widetilde{v}_{as} &= 3.4\widetilde{i}_{as} + j(377)(11 + 1.1) \times 10^{-3}\widetilde{i}_{as} + p(11 + 1.1) \times 10^{-3}\widetilde{i}_{as} \\
&= 3.4\widetilde{i}_{as} + j4.6\widetilde{i}_{as} + 0.012p\widetilde{i}_{as}
\end{aligned} \tag{2D-3}$$

When the system reaches steady state, $p\widetilde{i}_{as}$ becomes zero and (2D-3) becomes

$$\widetilde{v}_{as} = (3.4 + j4.6)\widetilde{i}_{as} \tag{2D-4}$$

In terms of steady-state phasor

$$\sqrt{2}\widetilde{V}_{as} = (3.4 + j4.6)\sqrt{2}\widetilde{I}_{as} \tag{2D-5}$$

or

$$\widetilde{V}_{as} = (3.4 + j4.6)\widetilde{I}_{as} \tag{2D-6}$$

SP2.4-1. Let $V_{as} = \sqrt{2}V_s \sin[\omega_e t + \theta_{esv}(0)]$ and $V_{bs} = -\sqrt{2}V_s \cos[\omega_e t + \theta_{esv}(0)]$ Express V_{qs}^e and V_{ds}^e. $[V_{qs}^e = \sqrt{2}V_s \sin\theta_{esv}(0),\ V_{ds}^e = \sqrt{2}V_s \cos\theta_{esv}(0)]$

SP2.4-2. Verify (2.4-8).

SP2.4-3. Determine $(\mathbf{K}_s)^{-1}$ for $\mathbf{K}_s$ for $\phi = \dfrac{\pi}{2}$ for i_{ds}.

$$\left[(\mathbf{K}_s)^{-1} = \begin{bmatrix} \cos\theta & -\sin\theta \\ \sin\theta & \cos\theta \end{bmatrix}; \mathbf{K}_s \neq (\mathbf{K}_s)^{-1} \right]$$

SP2.4-4. Express $\widetilde{f}_{bs}$ in terms of f_{qs}^e and f_{ds}^e. $[\widetilde{f}_{bs} = -j\left(f_{qs}^e - jf_{ds}^e\right)]$

SP2.4-5. Add $\widetilde{f}_{bs}$ to Fig. 2.4-1. $[-j\widetilde{f}_{as}]$

2.5   Problems

1   Repeat Example 2.A for the arbitrary reference variables rather than the synchronous reference frame.

2   Obtain (2.2-18) from (2.2-14) and (2.2-6).

3   Obtain (2.2-8) from (2.2-7).

4 Assume

$$I_{as} = \sqrt{2}\, I_s \cos\left[\omega_e t + \theta_{esi}(0)\right]$$
$$I_{bs} = -\sqrt{2}\, I_s \sin\left[\omega_e t + \theta_{esi}(0)\right]$$

Express mmf$_s$ for Fig. 2.2-2.

5 Determine the transformation to the arbitrary reference frame for Problem 4.

6 Verify (2.4-31).

7 Assume

$$v_{as} = V_a \cos \omega_e t$$
$$v_{bs} = V_b \sin \omega_e t$$

where $V_a \neq V_b$. Determine v_{qs} and v_{ds}

8 Express v_{as} and v_{bs} for current positive out of the machine.

9 Express mmf$_s$ for Problem 8.

10 Express $\tilde{v}_{as}$ for Problem 8.

References

1 N. Tesla, *My Inventions and Other Writings*, Penguin Books, 2011.

2 R. H. Park, "Two-Reaction Theory of Synchronous Machines – Generalize Method of Analysis –Part 1," *AIEE Trans.* Vol. 48, July 1929, pp. 716–727.

3 P. C. Krause, O. Wasynczuk, T. C. O'Connell, and M. Hasan, "Tesla's Contribution to Electric Machine Analysis," *2018 IEEE Power &Energy Society General Meeting (PESGM)*, 2018, pp. 1–1. doi: 10.1109/PESGM.2018.8586170.

4 P. C. Krause, *Reference Frame Theory*, IEEE Press, Wiley, Hoboken, NJ, USA,2020.

5 P. C. Krause, O. Wasynczuk, S. D. Pekarek, and T. C. O'Connell, *Electromechanical Motion Devices*, 3rd Ed. IEEE Press and Wiley, Hoboken, NJ, USA, 2020.

3

Symmetrical Induction Machine

3.1 Introduction

Although the induction machine is generally used as a means to convert electric power to mechanical work (motor action), it can also operate as a generator and convert mechanical work to electric power. Three-phase induction motors are commonly used in large-horsepower applications, for example, pump drives, steel mill drives, hoist drives, vehicle drives, and as a generator in wind turbine and low-head hydro applications. In applications where three-phase is not available or in low-power requirements as household applications, single-phase induction motors, which are covered in Chapter 7, are used. Single-phase induction motors develop torque in a manner similar to multiphase induction motors and can operate with direct connection of the stator windings to a single-phase wall outlet.

The analysis of symmetrical two- and three-phase induction machines is essentially the same. In addition, we will show that we can relate the two-pole machine to machines with any number of pole pairs. Therefore, we will consider the two-pole two-phase machine, since this enables us to become familiar with the theory and performance of induction machines without becoming inundated with three-phase trigonometric manipulations. When appropriate, we will point out the changes necessary to apply the work to a three-phase symmetrical induction machine once the derivation for the two-phase machine has been completed.

3.2 Symmetrical Machines

A disassembled four-pole two-phase $\frac{1}{10}$-hp 115-V induction motor, which is used in low-power control applications, is shown in Fig. 3.2-1. Also shown in Fig. 3.2-1 is the case that houses the speed-reduction gears for hospital bed application.

Introduction to Modern Analysis of Electric Machines and Drives, First Edition.
Paul C. Krause and Thomas C. Krause.
© 2023 The Institute of Electrical and Electronics Engineers, Inc.
Published 2023 by John Wiley & Sons, Inc.

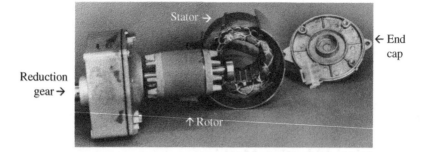

Figure 3.2-1 Four-pole two-phase 1/10-Hp 115-V induction motor with reduction gear.

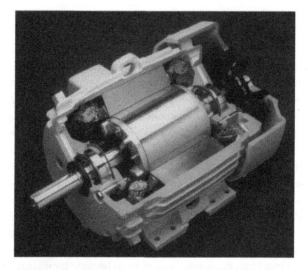

Figure 3.2-2 Four-pole three-phase 6.5-Hp 460-V severe-duty, squirrel-cage induction motor (courtesy of General Electric).

The rotors of the induction motors are referred to as "squirrel-cage" rotors. A cutaway of a four-pole three-phase 6.5-hp 460-V squirrel-cage induction motor is shown in Fig. 3.2-2. It is an enclosed fan-cooled severe-duty motor for use in the chemical, paper, cement, and mining industries. Although it is difficult to discern, the squirrel-cage rotors are made up of laminated punched steel with aluminum bars die casted in the openings of the laminated rotor and the bars terminated at each end of the rotor in an aluminum ring, which is visible in Figs. 3.2-1 and 3.2-2. The protrusions from the aluminum rings are for cooling purposes. If we remove the steel laminations, the remaining aluminum bars and end rings resemble the rotor (blades) of a "squirrel-cage fan." The question often arises as to how these short-circuited bars produce a rotating magnetic field. It turns out that these short-circuited bars behave like symmetrical windings. Short-circuited sinusoidally distributed windings are a good model for the rotor bars of a squirrel-cage induction machine. This requires considerable work to show the equivalence analytically; however, for now, let us assume that the rotor is equipped with symmetrical windings as shown in Fig. 3.2-3. Later we will show that this is a reasonable assumption.

For purposes of analysis, the two-pole two-phase induction machine is shown in Fig. 3.2-3. The stator windings are orthogonal, sinusoidally distributed windings as described in Chapter 2. We will also assume that the rotor of the two-pole two-phase induction machine may also be portrayed electrically by two sinusoidally distributed windings displaced 90°. Hence, for our present purposes, we will consider that the *ar* and *br* windings are sinusoidally distributed, each with the same number of turns and the same winding resistance. Thus, both the stator and rotor

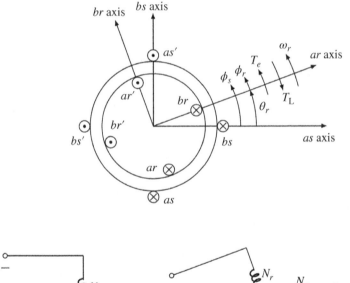

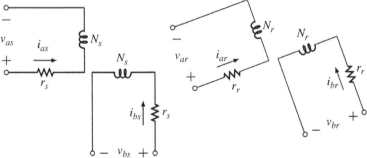

Figure 3.2-3 A two-pole two-phase, symmetrical machine. Note: $\phi_s = \theta_r + \phi_r$.

are symmetrical, that is, all stator windings have the same distribution and electrical parameters and all rotor windings have the same distribution and electrical parameters. For this reason, this device is often referred to as a *symmetrical machine* for analysis purposes.

Note that the air gap distance between the stator and rotor is uniform. The rotor windings of an induction machine are generally short-circuited ($v_{ar} = v_{br} = 0$) with only the stator windings connected to a source. In this case, the machine is said to be single-fed. In some special applications, such as wind turbines, both the rotor windings, which are coil-wound, and stator windings are connected to sources. In particular, the rotor windings are connected to a stationary multiphase source by a brush and slip-ring arrangement. In this case, the machine is double-fed. We will not deal with the double-fed machine.

As established in Chapter 2, the angular displacement about the stator is denoted ϕ_s, and it is referenced to the *as* axis. We see from Fig. 3.2-3 that the angular displacement about the rotor is denoted ϕ_r and it is referenced to the *ar* axis. The angular velocity of the rotor is ω_r and θ_r is its angular position between the *ar* and *as* axes. The electromechanical torque T_e and the load torque T_L are also indicated in Fig. 3.2-3. The torque T_e is assumed to be positive in the direction of increasing θ_r whereas the load torque T_L is positive in the opposite direction (opposing rotation).

Recall that Tesla's rotating magnetic field of the stator (mmf_s) rotates about the air gap at ω_e and the rotor windings see a changing flux linkage only when $\omega_r \neq \omega_e$. Since the currents induced in the rotor circuits are due to a time rate change of flux linkage, rotor currents would not be induced when the angular velocity of the rotor is equal to the angular velocity of the stator rotating magnetic field ($\omega_r = \omega_e$). Thus, the symmetrical induction machine will not develop a torque at synchronous speed; only when $\omega_r \neq \omega_e$ will rotor currents be induced and a torque produced. We will find that this is completely opposite to a synchronous machine which produces an average torque only when $\omega_r = \omega_e$. By now you have probably guessed that the induction machine is so named from the fact that torque is produced as a result the rotor currents being "induced" by the stator rotating magnetic field. It is interesting that the induction machine was first referred to as a rotating transformer. This is understandable and we will see this more clearly when we derive the steady-state equivalent circuit for the induction machine and find that it is similar to that of a transformer. Also, we should mention that generator action occurs if the stator windings are connected to a source, thus establishing a rotating magnetic field, and the rotor, with its short-circuited windings, is driven above synchronous speed.

SP3.2-1. The frequency of the balanced stator currents of the symmetrical machine is 60-Hz and mmf_s rotates counterclockwise. The device is operating as a motor, and the rotor of the two-pole machine is rotating counterclockwise at $0.9 \, \omega_e$. Determine the frequency of the rotor currents. [6-Hz]

SP3.2-2. Determine the direction of rotation of mmf$_s$ for (a) $\tilde{I}_b = -j\tilde{I}_a$ and (b) $\tilde{I}_b = j\tilde{I}_a$. [(a) ccw, (b) cw]

SP3.2-3. Assume sinusoidally distributed windings on the stator and rotor of the machine shown in Fig. 3.2-3. Express (a) mmf$_{as}$ in terms of θ_r and ϕ_r and (b) mmf$_{ar}$ in terms of θ_r and ϕ_s. [(a) mmf$_{as} = (N_s/2)i_{as}\cos(\phi_r + \theta_r)$; (b) mmf$_{ar} = (N_r/2)i_{ar}\cos(\phi_s - \theta_r)$]

3.3 Symmetrical Two-Pole Rotor Windings

3.3.1 Two-Phase Rotor Windings

The derivation of the air-gap mmf due to rotating symmetrical windings parallels that of stationary symmetrical windings. As shown in Fig. 3.3-1, the *ar* and *br* windings are orthogonal and if they are identical and sinusoidally distributed, Tesla's rotating magnetic field of these symmetrical windings may be expressed.

$$\text{mmf}_r = \frac{N_r}{2}(i_{ar}\cos\phi_r + i_{br}\sin\phi_r) \tag{3.3-1}$$

where N_r is the equivalent number of turns of the rotor windings and ϕ_r is its angular position from the *ar* axis, which is positioned θ_r from the *as* axis. In Fig. 3.3-1, the positions θ_r and θ, which locates the *q*-axis, are referenced to the *as* axis. The *q*-axis is the same *q*-axis introduced in Fig. 2.2-2.

Now, the positions are related as

$$\phi_r = \beta + \phi \tag{3.3-2}$$

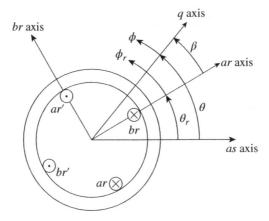

Figure 3.3-1 Two-phase rotating, identical, sinusoidally distributed windings.

Substituting (3.3-2) into (3.3-1) yields

$$\text{mmf}_r = \frac{N_r}{2}[i_{ar}\cos(\beta + \phi) + i_{br}\sin(\beta + \phi)] \tag{3.3-3}$$

We will return to (3.3-3) in a moment; however, let us continue our focus on the rotating air-gap mmf during balanced steady-state operation.

Now,

$$\beta = \theta - \theta_r \tag{3.3-4}$$

and during steady-state operation

$$\beta = (\omega - \omega_r)t + \theta(0) - \theta_r(0) \tag{3.3-5}$$

During balanced steady-state operation the currents, in the short-circuited rotor windings, which are induced by mmf_s, are balanced with the angular frequency of $(\omega_e - \omega_r)$. Thus,

$$I_{ar} = \sqrt{2}I_r\cos[(\omega_e - \omega_r)t + \theta_{eri}(0)] \tag{3.3-6}$$

$$I_{br} = \sqrt{2}I_r\sin[(\omega_e - \omega_r)t + \theta_{eri}(0)] \tag{3.3-7}$$

It is important to recall that I_{ar} and I_{br} are zero if $\omega_r = \omega_e$, so (3.3-6) and (3.3-7) are only valid for $\omega_e \neq \omega_r$.

If we now substitute (3.3-6), (3.3-7), and (3.3-4) into (3.3-3), we will obtain an expression for mmf_r:

$$\text{mmf}_r = \frac{N_r}{2}\sqrt{2}I_r\cos[(\omega_e - \omega_r - \omega + \omega_r)t + \theta_{eri}(0) - \theta(0) + \theta_r(0) - \phi] \tag{3.3-8}$$

Now, we will select $\theta(0)$ and $\theta_r(0)$ to be zero unless otherwise specified, whereupon, (3.3-8) may be written as

$$\text{mmf}_r = \frac{N_r}{2}\sqrt{2}I_r\cos[(\omega_e - \omega)t + \theta_{eri}(0) - \phi] \tag{3.3-9}$$

Equation (3.3-9) is similar in form to mmf_s given by (2.2-8). This is a very important observation. Recall from Chapter 2 that the rotating magnetic field produced by the stator comes about due to the sinusoidally distributed windings and the balanced stator currents. The two- and three-phase rotating magnetic fields differ only by a constant. The currents induced in the short-circuited rotor bars are unrestricted and one can argue that the stator mmf will induce rotor currents such that the resulting fundamental of the rotor rotating magnetic field (mmf_r) would be sinusoidal in waveform as the inducing mmf (mmf_s) with an angular velocity equal to the difference in rotor speed from synchronous speed. We see from (3.3-9) that this is what occurs if the rotor were equipped with short-circuited symmetrical sinusoidally distributed windings with currents of an angular frequency of $\omega_e - \omega_r$.

If the rotor currents are balanced and if the angular frequency of these currents is $(\omega_e - \omega_r)$, then the stator mmf_s and the rotor mmf_r will travel around the air gap at the same angular velocity regardless of the actual speed of the rotor if $\omega_r \neq \omega_e$. If, for example, we observe mmf_s and mmf_r as a stationary observer ($\omega = 0$), mmf_s^s and mmf_r^s would both be traveling counterclockwise at ω_e relative to us. If we are riding on the rotor ($\omega = \omega_r$), mmf_s^r and mmf_r^r would be traveling at $(\omega_e - \omega_r)$ counterclockwise relative to us. If we are running at ω_e, mmf_s^e and mmf_r^e would appear to us as constant space sinusoids of ϕ_e during the steady state. Since mmf_s and mmf_r are traveling in unison, a constant torque (power) is produced. The magnitude of the constant torque is determined by the equivalent number of turns of stator and rotor windings, machine geometry, the magnitude of the stator and rotor currents, and the relative position of mmf_s and mmf_r.

The relative position of the phase between the mmfs, $\theta_{esi}(0)$ and $\theta_{eri}(0)$, is easily discerned in the synchronous reference frame, i.e. mmf_s^e and mmf_r^e

$$mmf_s^e = \frac{N_s}{2} \sqrt{2} I_s \cos\left[\theta_{esi}(0) - \phi_e\right] \tag{3.3-10}$$

$$mmf_r^e = \frac{N_r}{2} \sqrt{2} I_r \cos\left[\theta_{eri}(0) - \phi_e\right] \tag{3.3-11}$$

A plot of mmf_s^e and mmf_r^e is shown in Fig. 3.3-2 for motor action.

A synchronous machine develops a constant torque only at synchronous speed, $\omega_r = \omega_e$. This occurs due to the fact that the rotor is either a permanent magnet or a field winding with a dc current which is stationary with respect to the rotor which

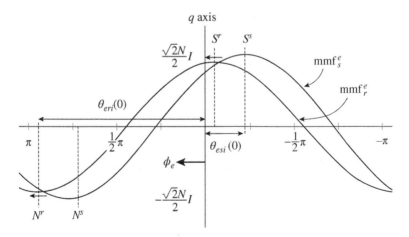

Figure 3.3-2 Plot of developed or unrolled view of mmf_s^e (3.3-10) and mmf_r^e (3.3-11), motor action.

must rotate in unison with Tesla's rotating magnetic field. This, of course, is necessary to produce an average torque. Now, the induction or symmetrical machine produces torque at any rotor speed except when $\omega_r = \omega_e$ and current is not induced in the short-circuited rotor circuits. In this case, mmf$_r$ disappears and mmf$_s$ has no other mmf with which to interact.

Let us take a minute to talk about the assumption we have made in regard to the rotor windings. As we have said earlier, the rotating mmf$_s$ induces currents in the rotor windings that will produce a rotor mmf$_r$ that is of the form as mmf$_s$. With the rotor windings assumed to be symmetrical windings, mmf$_r$ as viewed from the arbitrary reference frame (3.3-9) is of the same form as the stator rotating, mmf$_s$, (2.2-8). The assumption of rotor squirrel-cage windings as two short-circuited orthogonal symmetrical sinusoidally distributed windings appears justified. This is also true for the three-phase machine since the mmf$_s$ for the three-phase mmf differs from the two-phase only by a constant as given by (2.2-18).

3.3.2 Three-Phase Rotor Windings

This subsection is very similar to the derivation of the subsection entitled "Two-Pole Three-Phase Stator" in Section 2.2. For the three-phase rotor, mmf$_r$ differs from the two-phase by a constant $\frac{3}{2}$. In particular, in the arbitrary reference frame

$$\text{mmf}_r = \frac{N_r}{2}\sqrt{2}\,I_r\frac{3}{2}\cos\left[(\omega_e - \omega)t + \theta_{eri}(0) - \theta(0) - \phi\right] \tag{3.3-12}$$

In the synchronous rotating reference frame for $\theta(0) = 0$,

$$\text{mmf}_r^e = \frac{N_r}{2}\sqrt{2}\,I_r\frac{3}{2}\cos\left[\theta_{eri}(0) - \phi\right] \tag{3.3-13}$$

Therefore, it is not necessary to redo the derivation leading up to Fig. 3.3-2 since the three-phase case differs only by a constant from the two-phase case.

SP3.3-1. Let $\theta(0)$ be $\frac{\pi}{4}$ and express mmf$_s$ and mmf$_r$. Does this change the justification made for sinusoidal rotor windings? [No]

SP3.3-2. Write (3.3-6) and (3.3-7) for cw rotation mmf$_r$. $[I_{br} = -(3.3-7)]$

3.4 Substitute Variables for Symmetrical Rotating Circuits and Equivalent Circuit

3.4.1 Two-Phase Machine

The voltage equations expressed for two-phase stator circuits in the arbitrary reference frame are given by (2.4-11) and (2.4-12). We need similar equations for the symmetrical rotor circuits. The transformation for the rotor variables to the arbitrary reference frame follows that set forth in Section 2.4 for the stator circuits. In particular, if (3.3-2) is substituted into (3.3-1) and after a few trigonometric manipulations, we have

$$\text{mmf}_r = \frac{N_r}{2} \cos\phi(i_{ar}\cos\beta + i_{br}\sin\beta) + \frac{N_r}{2}\sin\phi(-i_{ar}\sin\beta + i_{br}\cos\beta) \quad (3.4\text{-}1)$$

If we let $\phi = 0$, we obtain i_{qr} and $\phi = -\dfrac{\pi}{2}$, we obtain i_{dr}. Equation (3.4-1) may now be written as

$$\text{mmf}_r = \frac{N_r}{2}\left(i_{qr}\cos\phi - i_{dr}\sin\phi\right) \quad (3.4\text{-}2)$$

Figure 3.4-1 depicts the fictitious qr and dr windings for the rotor. Please understand that the q- and d-axis are the same for the stator, Fig. 2.3-1, as for the rotor, Fig. 3.4-1. It is important to mention that if a device has symmetrical stator windings and symmetrical rotor windings, as is the case of an induction or symmetrical machine, the stator and rotor windings have symmetrical fictitious two-phase windings in the arbitrary reference frame. Moreover, we will find that in the arbitrary reference frame the stator and rotor fictitious windings are stationary relative to each other; therefore, position-dependent mutual inductances are eliminated in all reference frames.

The transformation for the rotating symmetrical circuits becomes

$$\begin{bmatrix} f_{qr} \\ f_{dr} \end{bmatrix} = \begin{bmatrix} \cos\beta & \sin\beta \\ \sin\beta & -\cos\beta \end{bmatrix} \begin{bmatrix} f_{ar} \\ f_{br} \end{bmatrix} \quad (3.4\text{-}3)$$

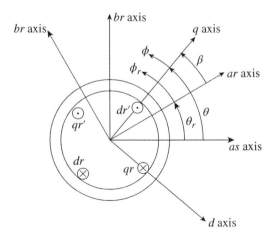

Figure 3.4-1 The fictitious two-phase qr and dr windings.

or

$$\mathbf{f}_{qdr} = \mathbf{K}_r \mathbf{f}_{abr} \qquad (3.4\text{-}4)$$

and

$$(\mathbf{K}_r)^{-1} = \mathbf{K}_r \qquad (3.4\text{-}5)$$

The voltages v_{ar} and v_{br} are similar in form to (2.4-1) and (2.4-2) with r rather than s in the subscript. By a procedure similar to that used for the stator voltages, the substitute arbitrary reference frame voltage equations for a resistive-inductive circuit become

$$v_{qr} = r_r i_{qr} + (\omega - \omega_r)\lambda_{dr} + p\lambda_{qr} \qquad (3.4\text{-}6)$$

$$v_{dr} = r_r i_{dr} - (\omega - \omega_r)\lambda_{qr} + p\lambda_{dr} \qquad (3.4\text{-}7)$$

These substitute voltage equations are very similar in form to those for stationary circuits given by (2.4-11) and (2.4-12), in particular, $\omega - \omega_r$ replaces ω.

We have two things to do before we can come up with equivalent circuits. In particular, we must refer the rotor variables to the stator windings and next we must express λ_{qds} and λ'_{qdr}. First, the turn ratios necessary to refer the rotor variables to the stator windings are

$$\mathbf{i}'_{abr} = \frac{N_r}{N_s} \mathbf{i}_{abr} \qquad (3.4\text{-}8)$$

$$\mathbf{v}'_{abr} = \frac{N_s}{N_r} \mathbf{v}_{abr} \qquad (3.4\text{-}9)$$

$$\lambda'_{abr} = \frac{N_s}{N_r} \lambda_{abr} \qquad (3.4\text{-}10)$$

The substitute variables may now be written in the arbitrary reference frame as

$$v_{qs} = r_s i_{qs} + \omega\lambda_{ds} + p\lambda_{qs} \qquad (3.4\text{-}11)$$

$$v_{ds} = r_s i_{ds} - \omega\lambda_{qs} + p\lambda_{ds} \qquad (3.4\text{-}12)$$

$$v'_{qr} = r'_r i'_{qr} + (\omega - \omega_r)\lambda'_{dr} + p\lambda'_{qr} \qquad (3.4\text{-}13)$$

$$v'_{dr} = r'_r i'_{dr} - (\omega - \omega_r)\lambda'_{qr} + p\lambda'_{dr} \qquad (3.4\text{-}14)$$

where

$$\mathbf{r}'_r = \left(\frac{N_s}{N_r}\right)^2 \mathbf{r}_r \qquad (3.4\text{-}15)$$

From Fig. 3.2-3, we can express the flux linkages as

$$\lambda_{as} = L_{asas}i_{as} + L_{asbs}i_{bs} + L_{asar}i_{ar} + L_{asbr}i_{br} \qquad (3.4\text{-}16)$$

$$\lambda_{bs} = L_{bsas}i_{as} + L_{bsbs}i_{bs} + L_{bsar}i_{ar} + L_{bsbr}i_{br} \qquad (3.4\text{-}17)$$

$$\lambda_{ar} = L_{aras}i_{as} + L_{arbs}i_{bs} + L_{arar}i_{ar} + L_{arbr}i_{br} \qquad (3.4\text{-}18)$$

$$\lambda_{br} = L_{bras}i_{as} + L_{brbs}i_{bs} + L_{brar}i_{ar} + L_{brbr}i_{br} \qquad (3.4\text{-}19)$$

The self- and mutual inductances are defined by their subscripts. Reciprocity applies, thus $L_{asbs} = L_{bsas}$, $L_{asar} = L_{aras}$, etc. For future derivations, it is convenient to write (3.4-16) through (3.4-19) in matrix form as

$$\begin{bmatrix} \lambda_{abs} \\ \lambda_{abr} \end{bmatrix} = \begin{bmatrix} \mathbf{L}_s & \mathbf{L}_{sr} \\ (\mathbf{L}_{sr})^T & \mathbf{L}_r \end{bmatrix} \begin{bmatrix} \mathbf{i}_{abs} \\ \mathbf{i}_{abr} \end{bmatrix} \qquad (3.4\text{-}20)$$

As in the case of the transformer, the self-inductance of each winding is made up of a leakage inductance caused by the flux that fails to cross the air gap and a magnetizing inductance due to the flux that traverses the air gaps and circulates through the stator and rotor steel. For symmetrical stator windings, the self-inductances L_{asas} and L_{bsbs} are equal and will be denoted L_{ss}, where

$$L_{ss} = L_{ls} + L_{ms} \qquad (3.4\text{-}21)$$

In (3.4-21), L_{ls} is the leakage inductance and L_{ms} the magnetizing inductance. The machine is designed to minimize the leakage inductance; it generally makes up approximately 10% of the self-inductance. The self-inductance of symmetrical rotor windings may be expressed similarly,

$$L_{rr} = L_{lr} + L_{mr} \qquad (3.4\text{-}22)$$

The magnetizing inductances L_{ms} and L_{mr} may be expressed in terms of turns and reluctance. In particular,

$$L_{ms} = \frac{N_s^2}{\mathcal{R}_m} \qquad (3.4\text{-}23)$$

$$L_{mr} = \frac{N_r^2}{\mathcal{R}_m} \qquad (3.4\text{-}24)$$

The magnetizing reluctance $\mathcal{R}_m$ is the reluctance seen by the magnetizing flux which includes the iron and both air gaps. It is dominated by the air gaps. Since the stator (rotor) windings are orthogonal and the rotor is round, coupling does not exist between the *as* and *bs* windings (L_{asas} or L_{bsbs}) or between the *ar* and *br* windings (L_{arbr} or L_{brar}). Recall that the equivalent, sinusoidally distributed windings are depicted by one coil placed at the maximum turn density. The mutual inductances L_{asbs}, L_{bsas}, L_{arbr}, and L_{brar} are all zero in the case of a two-phase stator. In a three-phase machine, where the stator (rotor) windings are displaced by 120° magnetically in space, a coupling exists between the stator (rotor) windings. Nevertheless, for the two-phase machine, we can write

$$\mathbf{L}_s = \begin{bmatrix} L_{ss} & 0 \\ 0 & L_{ss} \end{bmatrix} = L_{ss}\mathbf{I} \tag{3.4-25}$$

$$\mathbf{L}_r = \begin{bmatrix} L_{rr} & 0 \\ 0 & L_{rr} \end{bmatrix} = L_{rr}\mathbf{I} \tag{3.4-26}$$

Coupling will occur between the stator and rotor windings and this coupling will vary with the position (θ_r) of the rotor windings relative to the stator windings. For example, when the *as* and *ar* windings are aligned, $\theta_r = 0$, the magnitude of coupling between these windings is maximized and, with the assumed direction of positive i_{as} and i_{ar}, the right-hand rule tells us that the mutual fluxes are adding. Hence, the mutual inductance at $\theta_r = 0$ is a positive maximum and can be expressed in terms of turns and $\mathcal{R}_m$ as

$$L_{asar} = \frac{N_s N_r}{\mathcal{R}_m} \quad \text{for } \theta_r = 0 \tag{3.4-27}$$

Now, when $\theta_r = \frac{1}{2}\pi$, the *as* and *ar* windings are orthogonal and

$$L_{asar} = 0 \quad \text{for } \theta_r = \frac{1}{2}\pi \tag{3.4-28}$$

For $\theta_r = \pi$ the windings are again aligned but now, with the assumed direction of positive i_{as} and i_{ar}, they oppose, thus

$$L_{asar} = -\frac{N_s N_r}{\mathcal{R}_m} \quad \text{for } \theta_r = \pi \tag{3.4-29}$$

at $\theta_r = \frac{3}{2}\pi$, the windings are again orthogonal and

$$L_{asar} = 0 \quad \text{for } \theta_r = \frac{3}{2}\pi \tag{3.4-30}$$

From (3.4-27) through (3.4-30), we see that mutual inductances can be approximated as a cosine function of θ_r. In particular, if we define L_{sr} as

$$L_{sr} = \frac{N_s N_r}{\mathcal{R}_m} \tag{3.4-31}$$

we can write L_{asar} or L_{aras} as

$$L_{asar} = L_{sr} \cos \theta_r \tag{3.4-32}$$

If we were to carry out the derivation as in [1], we would find that (3.4-32) is, indeed, a valid expression for the mutual inductance between the *as* and *ar* sinusoidally distributed windings. It follows by inspection of Fig. 3.2-3 that

$$L_{asbr} = -L_{sr} \sin \theta_r \tag{3.4-33}$$

$$L_{bsar} = L_{sr} \sin \theta_r \tag{3.4-34}$$

$$L_{bsbr} = L_{sr} \cos \theta_r \tag{3.4-35}$$

Hence,

$$\mathbf{L}_{sr} = L_{sr} \begin{bmatrix} \cos \theta_r & -\sin \theta_r \\ \sin \theta_r & \cos \theta_r \end{bmatrix} \tag{3.4-36}$$

Once the expressions for the mutual inductances are known, we begin to under-stand the complexities involved in the analysis of electric machines. The stator-to-rotor mutual inductances are sinusoidal functions of θ_r because of the relative motion. In the voltage equations, we take the derivative of the flux linkages with respect to time, we no longer obtain only the familiar $L(di/dt)$ term. Instead, two terms result; one due to the derivative of the mutual inductance, since θ_r is a func-tion of time, and one due to the derivative of the current. For example,

$$\frac{d(L_{asar}i_{as})}{dt} = \frac{\partial(L_{asar}i_{as})}{\partial \theta_r} \frac{d\theta_r}{dt} + \frac{\partial(L_{asar}i_{as})}{\partial i_{as}} \frac{di_{as}}{dt} \tag{3.4-37}$$

Substitution of (3.4-8) and (3.4-10) into (3.4-20) yields

$$\begin{bmatrix} \boldsymbol{\lambda}_{abs} \\ \boldsymbol{\lambda}'_{abr} \end{bmatrix} = \begin{bmatrix} \mathbf{L}_s & \dfrac{N_s}{N_r}\mathbf{L}_{sr} \\ \dfrac{N_s}{N_r}(\mathbf{L}_{sr})^T & \mathbf{L}'_r \end{bmatrix} \begin{bmatrix} \mathbf{i}_{abs} \\ \mathbf{i}'_{abr} \end{bmatrix} \tag{3.4-38}$$

where

$$\mathbf{L}'_r = \left(\frac{N_s}{N_r}\right)^2 \mathbf{L}_r = \begin{bmatrix} L'_{rr} & 0 \\ 0 & L'_{rr} \end{bmatrix} \tag{3.4-39}$$

Since L_{mr} and L_{ms} may be related from (3.4-23) and (3.4-24) and $L'_{lr} = \left(\dfrac{N_s}{N_r}\right)^2 L_{lr}$,

$$L'_{rr} = L'_{lr} + \left(\frac{N_s}{N_r}\right)^2 L_{mr} \tag{3.4-40}$$

$$= L'_{lr} + L_{ms}$$

Note that

$$\frac{N_s}{N_r}\mathbf{L}_{sr} = \frac{N_s}{N_r}L_{sr}\begin{bmatrix} \cos \theta_r & -\sin \theta_r \\ \sin \theta_r & \cos \theta_r \end{bmatrix} \tag{3.4-41}$$

Comparing L_{ms} and L_{sr}, we see that

$$\frac{N_s}{N_r}L_{sr} = L_{ms} \tag{3.4-42}$$

Hence, (3.4-41) may be expressed in terms of L_{ms} and, for compactness, we will define $\mathbf{L}'_{sr}$ as

$$\mathbf{L}'_{sr} = \frac{N_s}{N_r}\mathbf{L}_{sr}$$

$$= L_{ms}\begin{bmatrix} \cos\theta_r & -\sin\theta_r \\ \sin\theta_r & \cos\theta_r \end{bmatrix} \tag{3.4-43}$$

Thus, (3.4-38) becomes

$$\begin{bmatrix} \boldsymbol{\lambda}_{abs} \\ \boldsymbol{\lambda}'_{abr} \end{bmatrix} = \begin{bmatrix} \mathbf{L}_s & \mathbf{L}'_{sr} \\ (\mathbf{L}'_{sr})^T & \mathbf{L}'_r \end{bmatrix}\begin{bmatrix} \mathbf{i}_{abs} \\ \mathbf{i}'_{abr} \end{bmatrix} \tag{3.4-44}$$

The flux linkage equations for a magnetically linear system given by (3.4-44) are in terms of as, bs, ar', and br' variables. In the arbitrary reference frame, (3.4-44) becomes

$$\begin{bmatrix} \boldsymbol{\lambda}_{qds} \\ \boldsymbol{\lambda}'_{qdr} \end{bmatrix} = \begin{bmatrix} \mathbf{K}_s\mathbf{L}_s(\mathbf{K}_s)^{-1} & \mathbf{K}_s\mathbf{L}'_{sr}(\mathbf{K}_r)^{-1} \\ \mathbf{K}_r(\mathbf{L}'_{sr})^T(\mathbf{K}_s)^{-1} & \mathbf{K}_r\mathbf{L}'_r(\mathbf{K}_r)^{-1} \end{bmatrix}\begin{bmatrix} \mathbf{i}_{qds} \\ \mathbf{i}'_{qdr} \end{bmatrix} \tag{3.4-45}$$

Since, $\mathbf{L}_s = L_{ss}\mathbf{I}$ and $\mathbf{L}'_r = L'_{rr}\mathbf{I}$,

$$\mathbf{K}_s\mathbf{L}_s(\mathbf{K}_s)^{-1} = \begin{bmatrix} L_{ss} & 0 \\ 0 & L_{ss} \end{bmatrix} \tag{3.4-46}$$

$$\mathbf{K}_r\mathbf{L}'_r(\mathbf{K}_r)^{-1} = \begin{bmatrix} L'_{rr} & 0 \\ 0 & L'_{rr} \end{bmatrix} \tag{3.4-47}$$

Now let us work on the upper right element of the 2 × 2 matrix in (3.4-45)

$$\mathbf{K}_s\mathbf{L}'_{sr}(\mathbf{K}_r)^{-1} = \begin{bmatrix} \cos\theta & \sin\theta \\ \sin\theta & -\cos\theta \end{bmatrix} L_{ms}\begin{bmatrix} \cos\theta_r & -\sin\theta_r \\ \sin\theta_r & \cos\theta_r \end{bmatrix}\begin{bmatrix} \cos(\theta-\theta_r) & \sin(\theta-\theta_r) \\ \sin(\theta-\theta_r) & -\cos(\theta-\theta_r) \end{bmatrix} \tag{3.4-48}$$

It is left to the reader to show that (3.4-48) reduces to

$$\mathbf{K}_s\mathbf{L}'_{sr}(\mathbf{K}_r)^{-1} = \begin{bmatrix} L_{ms} & 0 \\ 0 & L_{ms} \end{bmatrix} \tag{3.4-49}$$

and that

$$\mathbf{K}_s\mathbf{L}'_{sr}(\mathbf{K}_r)^{-1} = \mathbf{K}_r(\mathbf{L}'_{sr})^T(\mathbf{K}_s)^{-1} \tag{3.4-50}$$

Therefore, due to the fact that L_{ms}, L_{mr}, and L_{sr} have a common term, $\mathcal{R}_m$, we can write

$$\begin{bmatrix} \lambda_{qds} \\ \lambda'_{qdr} \end{bmatrix} = \begin{bmatrix} \mathbf{L}_s & \mathbf{L}_{ms} \\ \mathbf{L}_{ms} & \mathbf{L}'_r \end{bmatrix} \begin{bmatrix} \mathbf{i}_{qds} \\ \mathbf{i}'_{qdr} \end{bmatrix} \tag{3.4-51}$$

Since $L_{ss} = L_{ls} + L_{ms}$ and $L'_{rr} = L'_{lr} + L_{ms}$, (3.4-51) may be expressed as

$$\lambda_{qs} = L_{ls}i_{qs} + L_{ms}\left(i_{qs} + i'_{qr}\right) \tag{3.4-52}$$

$$\lambda_{ds} = L_{ls}i_{ds} + L_{ms}\left(i_{ds} + i'_{dr}\right) \tag{3.4-53}$$

$$\lambda'_{qr} = L'_{lr}i'_{qr} + L_{ms}\left(i_{qs} + i'_{qr}\right) \tag{3.4-54}$$

$$\lambda'_{dr} = L'_{lr}i'_{dr} + L_{ms}\left(i_{ds} + i'_{dr}\right) \tag{3.4-55}$$

Equations (3.4-11) through (3.4-14) and (3.4-52) through (3.4-55) suggest the equivalent circuits given in Fig. 3.4-2. Equations (3.4-52) through (3.4-55) and Fig. 3.4-2 are important in that for a symmetrical induction machine, the rotor position-dependent terms of the mutual inductances are eliminated in all frames of reference. We would expect this since both the stator and rotor windings are

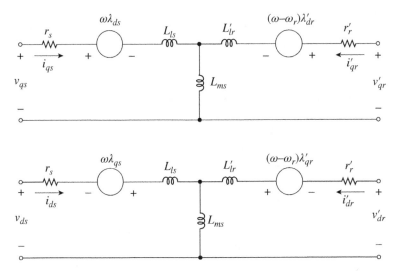

Figure 3.4-2 Arbitrary reference frame equivalent circuits for a two-phase, symmetrical induction machine.

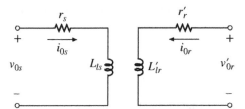

Figure 3.4-3 Zero-variables equivalent circuits to be added to Fig. 3.4-2 for a three-phase machine.

assumed to be symmetrical and we are able to transform the variables to a common frame of reference, the arbitrary reference frame.

3.4.2 Three-Phase Machine

As a result of Park's $\frac{2}{3}$ factor in the transformation of three-phase variables, the substitute variables are the same form for the two-phase and three-phase systems. Also, due to the fact that most stator and rotor windings are Y-connected, the sum of the stator and rotor currents is zero; therefore, L_{ms} is replaced with L_{Ms}. Thus, Fig. 3.4-2 is valid for a three-phase machine if L_{ms} is replaced with L_{Ms} and if the equivalent circuits, shown in Fig. 3.4-3, for the zero-variables, which are zero for balanced conditions and not a function of θ_r or θ, are added to Fig. 3.4-2.

SP3.4-1. If in the equivalent circuits given in Fig. 3.4-2, there are speed voltages only in the qr' and dr' part of the T equivalent circuits, what reference frame is being used? [stationary]

SP3.4-2. Repeat SP3.4-1 if speed voltages are only in the qs and ds part. [rotor]

SP3.4-3. Redraw Fig. 3.4-2 for a blocked rotor test of a squirrel-cage rotor. [ω_r and ω both zero, $v^s_{qr} = v^s_{dr} = 0, v^s_{qs} = v_{as}, v^s_{ds} = -v_{bs}$]

3.5 Electromagnetic Force and Torque

Our goal in this section is to derive an expression for the electromagnetic torque developed in rotational systems or force in translational systems. Although there are several approaches that could be used for this derivation, with the background that we have established, the "energy balance" approach is perhaps most convenient [2]. This method is quite direct and results in an easy-to-use expression for evaluating torque or force in electromagnetic systems.

Electromechanical devices consist of an electrical system, a coupling field, and a mechanical system. The electric and the mechanical systems can either supply or absorb energy by way of the coupling field which can either be magnetic, which will be our focus, or electric. The coupling field stores energy and transfers energy

Figure 3.5-1 Block diagram of possible energy interchange in an elementary electromechanical system.

between the two systems. A block diagram depicting this interaction and possible directions of energy flow is given in Fig. 3.5-1.

The energy balance is shown in Fig. 3.5-2. W_E and W_M are positive for energy being supplied to the electromechanical device from an external electrical system (W_E) and mechanical system (W_M). Likewise, W_e and W_m are positive when supplying energy to the coupling field in the electrical form (W_e) and mechanical form (W_m).

In Fig. 3.5-2, the subscripts E and e pertain to the electrical system, M and m to the mechanical, and f to the field. Also, the subscripts S and L indicate energy stored and energy lost, respectively. The energy W_{eL} is the energy lost due to resistive losses (i^2r); W_{eS} is the energy stored in a field external to the coupling field. Some authors consider energy stored in the field of the leakage inductances as W_{eS} while others will include the leakage flux in the coupling field. This is irrelevant since the energy stored in the field of the leakage inductance is generally not a function of the mechanical motion of the electromechanical device.

The energy W_{mL} represents energy lost due to mechanical friction and windage losses. W_{mS} is the energy stored as kinetic energy in the rotating mass (rotor) or as potential energy in a spring.

The energy W_e is the energy coming into the coupling field from the electric system. The energy W_m is the energy coming into the coupling field from the mechanical system. The energy W_{fL} is the energy lost due to hysteresis loss and circulating currents induced in the mechanical components of the coupling field. Electric machines are designed to minimize these losses making the λi relationship

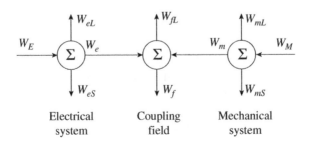

Figure 3.5-2 Energy balance.

approach a single-valued function as shown in Fig. 1.3-3. We will neglect W_{fL}, whereupon, we can express the energy balance of the coupling field as

$$W_f = W_e + W_m \tag{3.5-1}$$

where W_f is the energy stored in the coupling field.

The torque relationship between the electromechanical device and the mechanical system may be written as

$$T_e = J\frac{d\omega_r}{dt} + B_m\omega_r + T_L \tag{3.5-2}$$

which is the rotational form of Newton's second law. Here, J is the inertia of the rotor and any tightly mechanically coupled rotating mass, B_m, is the mechanical damping which is generally small and often neglected. Note that positive T_e acts to increase ω_r whereas a positive load torque T_L would act to retard ω_r. Therefore, with T_e positive for motor action, then positive W_m entering the coupling field as shown in Fig. 3.5-2 and as expressed in (3.5-1), would be

$$W_m = -\int T_e d\theta_r \tag{3.5-3}$$

The energy entering the coupling field from the two electrical sources shown in Fig. 3.2-3 W_e is

$$W_e = \int \left(i_1\frac{d\lambda_1}{dt}dt + i_2\frac{d\lambda_2}{dt}dt \right) \tag{3.5-4}$$

where the subscripts 1 and 2 are as and bs, respectively.

Therefore, (3.5-4) may be written as

$$W_e = \int (i_{as}d\lambda_{as} + i_{bs}d\lambda_{bs}) \tag{3.5-5}$$

For J electrical inputs, (3.5-5) may be expressed as

$$W_e = \int \sum_{j=1}^{J} i_j d\lambda_j \tag{3.5-6}$$

where in the case of a machine, J is the number of stator phases and where j is an index $(1,2,3,...\,J)$ not to be confused with the imaginary part of a complex number or with inertia or Joules. Thus, (3.5-1) may now be written as

$$W_f = \int \sum_{j=1}^{J} i_j d\lambda_j - \int T_e d\theta_r \tag{3.5-7}$$

In differential form, which we will use extensively, (3.5-7) can be written as

$$T_e d\theta_r = \sum_{j=1}^{J} i_j d\lambda_j - dW_f \qquad (3.5\text{-}8)$$

From (1.3-38) for multiple electrical inputs

$$\sum_{j=1}^{J} \lambda_j i_j = W_c + W_f \qquad (3.5\text{-}9)$$

where W_c is the coenergy. If we take the derivative of (3.5-9), we have

$$\sum_{j=1}^{J} \lambda_j di_j + \sum_{j=1}^{J} i_j d\lambda_j = dW_c + dW_f \qquad (3.5\text{-}10)$$

Solving (3.5-10) for $\sum_{j=1}^{J} i_j d\lambda_j$ and substituting the result into (3.5-8) yields

$$T_e d\theta_r = -\sum_{j=1}^{J} \lambda_j di_j + dW_c \qquad (3.5\text{-}11)$$

Although (3.5-8) and (3.5-11) are valid for magnetically linear and nonlinear systems, we will consider only magnetically linear systems. Therefore, it is convenient to express the flux linkages in terms of currents and either θ_r for rotational systems or x for translational systems. Hence, we will choose currents and θ_r or x as independent variables. In this case, (3.5-11) is most convenient for obtaining an expression for T_e which can be written as

$$T_e(\mathbf{i}, \theta_r) d\theta_r = -\sum_{j=1}^{J} \lambda_j(\mathbf{i}, \theta_r)\, di_j + dW_c(\mathbf{i}, \theta_r) \qquad (3.5\text{-}12)$$

for compactness,

$$\mathbf{i} = (i_1, i_2, i_3, ..., i_J) \qquad (3.5\text{-}13)$$

Equation (3.5-12) may now be written as

$$T_e(\mathbf{i}, \theta_r) d\theta_r = -\sum_{j=1}^{J} \lambda_j(\mathbf{i}, \theta_r) di_j + \sum_{j=1}^{J} \frac{\partial W_c}{\partial i_j}(\mathbf{i}, \theta_r) di_j + \frac{\partial W_c(\mathbf{i}, \theta_r)}{\partial \theta_r} d\theta_r \qquad (3.5\text{-}14)$$

Equating coefficients of $d\theta_r$,

$$T_e(\mathbf{i}, \theta_r) = \frac{\partial W_c}{\partial \theta_r} \qquad (3.5\text{-}15)$$

In case of a translational system

$$f_e(\mathbf{i}, x) = \frac{\partial W_c}{\partial x} \tag{3.5-16}$$

where f_e is the electromagnetic force and x is the displacement. Since we will consider only magnetically linear systems $W_c = W_f$.

Example 3.A   Force Between Pole Faces of Air Gap

Determine the force that exists between the pole faces of the elementary magnetic circuit given in Fig. 1.3-1. The parameters of the magnetic circuit is shown in Fig. 1.3-1 are: $r = 1\,\Omega$, $N = 100$ turns, $v = 10\,\mathrm{V}$, $\ell_i = 40\,\mathrm{cm}$, $x = 3\,\mathrm{mm}$, $A_i = A_g = 40\,\mathrm{cm}^2$, $\mu_0 = 4\pi \times 10^{-7}$ H/m, and $\mu_{ri} = 1000$.

In order to make use of (3.5-16) to calculate the electromagnetic force f_e, we must allow what is referred to as a virtual displacement. That is, we will assume that the air-gap length x increases in the positive direction by dx; therefore, according to (3.5-16), f_e will be positive if it acts to lengthen the air gap. If f_e is negative, it acts to shorten the air gap.

From (1.3-9), the self-inductance is

$$L = L_l + L_m \tag{3A-1}$$

where L_l is the leakage inductance and L_m is the magnetizing inductance. In particular,

$$L = \frac{N^2}{\mathcal{R}_l} + \frac{N^2}{\mathcal{R}_i + \mathcal{R}_g} \tag{3A-2}$$

The coenergy is

$$W_c = \frac{1}{2}(L_l + L_m)i^2 \tag{3A-3}$$

In order to determine f_e, we must take the partial derivative of W_c with respect to x. Since L_l is not a function of x, we need only concern ourselves with L_m. We can write L_m as

$$
\begin{aligned}
L_m &= \frac{N^2}{\mathcal{R}_i + \mathcal{R}_g} \\
&= \frac{N^2}{\dfrac{\ell_i}{\mu_{ri}\mu_0 A_i} + \dfrac{x}{\mu_0 A_g}}
\end{aligned} \tag{3A-4}
$$

which can be written as

$$L_m(x) = \frac{k_0}{k_1 + k_2 x}$$

$$= \frac{\dfrac{k_0}{k_2}}{\dfrac{k_1}{k_2} + x} \tag{3A-5}$$

where

$$k_0 = N^2$$
$$= (1 \times 10^2)^2 = 1 \times 10^4 \text{ turns} \tag{3A-6}$$

$$k_1 = \frac{\ell_i}{\mu_{ri}\mu_0 A_i}$$
$$= \frac{4 \times 10^{-1}}{(1 \times 10^3)(4\pi \times 10^{-7})(4 \times 10^{-4})} = 7.96 \times 10^5 \text{ H}^{-1} \tag{3A-7}$$

$$k_2 = \frac{1}{\mu_0 A_g}$$
$$= \frac{1}{(4\pi \times 10^{-7})(4 \times 10^{-4})} = 1.99 \times 10^9 \text{ m/H} \tag{3A-8}$$

Therefore,

$$W_c = \frac{\dfrac{1}{2}\dfrac{k_0}{k_2}i^2}{\dfrac{k_1}{k_2} + x} \tag{3A-9}$$

Now, from (3.5-16),

$$f_e(i, x) = \frac{\partial W_c}{\partial x}$$

$$= -\frac{\dfrac{1}{2}\left(\dfrac{k_0}{k_2}i^2\right)}{\left(\dfrac{k_1}{k_2} + x\right)^2}$$

$$= -\frac{\dfrac{1}{2}(5.025 \times 10^{-6} \times 10^2)}{(3.995 \times 10^{-4} + 3 \times 10^{-3})^2} = -21.74 \text{ N (an attractive force)} \tag{3A-10}$$

A force is established in a magnetic system to minimize the reluctance.

Let us now consider the two-phase symmetrical machine. From (3.4-25), (3.4-39), and (3.4-44), we can express the field energy as

$$W_f(\mathbf{i}, \theta_r) = \frac{1}{2}L_{ss}i_{as}^2 + \frac{1}{2}L_{ss}i_{bs}^2 + \frac{1}{2}L'_{rr}i'^2_{ar} + \frac{1}{2}L'_{rr}i'^2_{br} + L_{ms}i_{as}i'_{ar}\cos\theta_r - L_{ms}i_{as}i'_{br}\sin\theta_r$$
$$+ L_{ms}i_{bs}i'_{ar}\sin\theta_r + L_{ms}i_{bs}i'_{br}\cos\theta_r$$

$$(3.5\text{-}17)$$

For a magnetically linear system $W_f = W_c$ and T_e becomes

$$T_e = -L_{ms}\left[\left(i_{as}i'_{ar} + i_{bs}i'_{br}\right)\sin\theta_r + \left(i_{as}i'_{br} - i_{bs}i'_{ar}\right)\cos\theta_r\right] \qquad (3.5\text{-}18)$$

Now since $\theta_r = \theta - \beta$, and with considerable work, (3.5-18) may be written in terms of substitute variables for a two-pole machine, as

$$T_e = L_{ms}\left(i_{qs}i'_{dr} - i_{ds}i'_{qr}\right) \qquad (3.5\text{-}19)$$

where T_e is positive for motor action. Also, (3.5-19) can be written as

$$T_e = \lambda'_{qr}i'_{dr} - \lambda'_{dr}i'_{qr} \qquad (3.5\text{-}20)$$

or

$$T_e = \lambda_{ds}i_{qs} - \lambda_{qs}i_{ds} \qquad (3.5\text{-}21)$$

We can determine the torque by a different approach. That is, the torque is the cross product of the total flux, ϕ, and the total mmf. To do this we will start with mmf$_s$ and mmf$_r$ in the arbitrary reference frame from (2.3-7) and (3.4-2):

$$\text{mmf}_s = \frac{N_s}{2}\left(i_{qs}\cos\phi + i_{ds}\sin\phi\right) \qquad (3.5\text{-}22)$$

$$\text{mmf}_r = \frac{N_r}{2}\left(i_{qr}\cos\phi + i_{dr}\sin\phi\right) \qquad (3.5\text{-}23)$$

These expressions are for one half of the mmfs, so that the total MMFs would be

$$\text{MMF}_s = N_s\left(i_{qs}\cos\phi + i_{ds}\sin\phi\right) \qquad (3.5\text{-}24)$$

$$\text{MMF}'_r = N_s\left(i'_{qr}\cos\phi + i'_{dr}\sin\phi\right) \qquad (3.5\text{-}25)$$

where MMF$'_r$ is referred to the N_s winding. Now, in order to obtain the total flux, we must divide (3.5-24) by the total reluctance which includes both air gaps. The cross product, T_e, becomes

$$T_e = \frac{\text{MMF}_s}{\mathcal{R}_m} \times \text{MMF}'_r$$
$$= \frac{N_s}{\mathcal{R}_m}\left(i_{qs}\cos\phi + i_{ds}\sin\phi\right) \times N_s\left(i'_{qr}\cos\phi + i'_{dr}\sin\phi\right)$$

$$(3.5\text{-}26)$$

which becomes

$$T_e = L_{ms} \left(i_{qs} i'_{dr} - i_{ds} i'_{qr} \right)$$ (3.5.27)

which is (3.5.19). For the three-phase machine, we need only to replace L_{ms} in (3.5-19), (3.5-20), (3.5-21), and (3.5-27) with L_{Ms}, and multiply (3.5.27) by $\frac{3}{2}$ due to Park's $\frac{2}{3}$ factor.

From (3.5-19) and (3.5.27), torque can be evaluated by the same variables in any reference frame. In other words, torque is invariant regardless of the reference frame. Another way of looking at this is the input power may be expressed as

$$P = (\mathbf{V}_{abs})^T i_{abs}$$ (3.5-28)

In terms of qs- and ds-variables,

$$(\mathbf{V}_{abs})^T i_{abs} = \left[(\mathbf{K}_s)^{-1} \mathbf{v}_{qds} \right] \mathbf{K}_s i_{qds}$$
$$= \left(\mathbf{v}_{qds} \right)^T - \left[(\mathbf{K}_s)^{-1} \right]^T (\mathbf{K}_s)^{-1} i_{qds}$$ (3.5-29)

Now,

$$\left[(\mathbf{K}_s)^{-1} \right]^T = \mathbf{K}_s$$ (3.5-30)

Therefore,

$$\left(\mathbf{v}_{abs} \right)^T i_{abs} = \left(\mathbf{v}_{qds} \right)^T i_{qds}$$ (3.5-31)

Now, $T_e \omega_r$ is the power output, therefore, since power input is invariant in all reference frames and since ω_r is a scalar, T_e is also invariant in all reference frames.

SP3.5-1. Neglect the reluctance of the ferromagnetic core in Example 3.A and calculate f_e. [−27.92 N]

SP3.5-2. Rearrange the magnetic circuit in Fig. 1.3-1 to produce a repelling force between pole faces. [The force will always be in the direction to minimize the reluctance of the magnetic system.]

SP3.5-3. Using mmf$_s$ and mmf$_r$ obtain T_e for $\phi_s = \frac{\pi}{2}$ for i_{ds} and following the procedure used for (3.5-26).

SP3.5-4. Obtain (3.5.27) from (3.5-26). [$\cos \phi \times \cos \phi = 0, \cos \phi \times \sin \phi = 1$, $\sin \phi \times \cos 0 = -1$]

3.6 P-Pole Machines

Thus far, we have considered only two-pole electromechanical motion devices; however, they may have any even number of pole pairs (2, 4, 6, 8, ...) up to more than 40 in the case of large hydroturbine generators. It happens that with a simple change of variables we can analyze all machines as if they were two-pole devices. We need only to modify the expression for evaluating torque and the actual rotor speed of a machine with more than two poles will be a multiple less than a two-pole machine.

The rotating air-gap mmf of a machine with more than two poles can be determined by considering the four-pole device shown in Fig. 3.6-1. In this figure, each phase winding consists of two series-connected windings, each of which is assumed to be sinusoidally distributed. For example, $as1'$ represents a group of conductors sinusoidally distributed over $0 < \phi_s < \frac{1}{2}\pi$. The phase windings consist

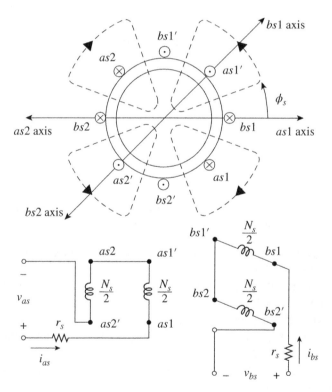

Figure 3.6-1 Stator winding arrangement of a four-pole, two-phase symmetrical electromechanical device. Flux streamlines shown for I_{as} maximum positive and I_{bs} is zero.

of N_s turns, with $N_s/\frac{P}{2}$ turns in each of the two series-connected windings. There may be some confusion in regard to notation. In Fig. 3.6-1, the notation as_1, bs_1, ... is used to denote sinusoidally distributed windings; however, a subscripted number denotes a coil.

Note that the maximum turns density for a winding occurs every $\frac{1}{2}\pi$ radians and each phase winding establishes two magnetic systems. As shown in Fig. 3.6-1, at the instant when $I_{bs} = 0$ and I_{as} is a positive maximum, the $as_1 - as_1'$ part of the as winding produces positive flux in the as_1-axis direction whereas the $as_2 - as_2'$ part of the as winding produces positive flux in the as_2-axis direction. South poles occur from $\phi_s = -\frac{1}{4}\pi$ to $\frac{1}{4}\pi$ and $\phi_s = \frac{3}{4}\pi$ to $\frac{5}{4}\pi$. Now, as shown in Fig. 3.6-1 by the streamlines, half of the flux that enters the stator steel from $-\frac{1}{4}\pi < \phi_s < \frac{1}{4}\pi$ reenters the air gap between $\phi_s = \frac{\pi}{4}$ and $\frac{\pi}{2}$; the other half between $\phi_s = -\frac{\pi}{4}$ and $-\frac{\pi}{2}$. The flux that enters the stator from $\phi_s = \frac{3}{4}\pi$ to $\frac{5}{4}\pi$ divides similarly. Hence, for $I_{as} = \sqrt{2}I_s$ and $I_{bs} = 0$, two north poles occur; one from $\phi_s = \frac{\pi}{4}$ to $\phi_s = \frac{3}{4}\pi$ and from $\phi_s = \frac{5}{4}\pi$ to $\phi_s = \frac{7}{4}\pi$.

The air-gap mmf established by each phase is a sinusoidal function of $2\phi_s$ for a four-pole machine, or, in general, $(P/2)\phi_s$, where P is the number of poles. In particular,

$$\text{mmf}_{as} = \frac{2}{P}\frac{N_s}{2}i_{as}\cos\frac{P}{2}\phi_s \tag{3.6-1}$$

$$\text{mmf}_{bs} = \frac{2}{P}\frac{N_s}{2}i_{bs}\sin\frac{P}{2}\phi_s \tag{3.6-2}$$

where N_s is the total equivalent turns per stator phase. For balanced steady-state operation, the stator currents may be expressed as

$$I_{as} = \sqrt{2}I_s\cos[\omega_e t + \theta_{esi}(0)] \tag{3.6-3}$$

$$I_{bs} = \sqrt{2}I_s\sin[\omega_e t + \theta_{esi}(0)] \tag{3.6-4}$$

Equations (3.6-3) and (3.6-4) are the same expressions for the stator currents as given for the two-pole case. Also, (2.2-9) would become

$$\text{mmf}_s^s = \left(\frac{2}{P}\right)\frac{N_s}{2}\sqrt{2}I_s\cos\left[\omega_e t + \theta_{esi}(0) - \frac{P}{2}\phi_s\right] \tag{3.6-5}$$

With the stator arranged as in Fig. 3.6-1, balanced steady-state stator currents of frequency ω_e produce a four-pole (P-pole) magnetic system that rotates about the air gap at $(2/4)\omega_e$ or $(2/P)\omega_e$ relative to a stationary observer. How do we know that? Well in order for (3.6-5) to be a constant, the argument of the cosine must be a constant and this occurs when $\dfrac{P}{2}\phi_s$ is equal to $\omega_e t$. In other words, when $\phi_s = \dfrac{2}{P}\omega_e t$. Therefore, the speed of the rotating mmf (rotating poles) is now $(2/P)\omega_e$ which is the synchronous speed of the machine; however, the stator variables are unaware of this. To the electric system, ω_e is synchronous speed.

A four-pole stator winding and a four-pole rotor winding are shown in Fig. 3.6-2. The mutual inductance between the *as-* and *ar*-windings may be expressed as

$$L_{asar} = L_{sr}\cos 2\theta_{rm} \tag{3.6-6}$$

where θ_{rm} is the actual rotor displacement. In the two-pole case, Fig. 2.2-1, we used θ_r. We are going to give a new meaning to θ_r in just a minute. Let us first generalize the argument of the cosine in (3.6-6). Thus,

$$L_{asar} = L_{sr}\cos \frac{P}{2}\theta_{rm} \tag{3.6-7}$$

If we let

$$\frac{P}{2}\theta_{rm} = \theta_r \tag{3.6-8}$$

then (3.6-7) becomes what we had for the two-pole case. Now, θ_{rm} is the actual rotor displacement and θ_r is what is referred to as the electrical angular position of the rotor. What in the world does that mean? Well, let us be stator variables; we do not know or care how many pole pairs the machine has, so we will assume that both the stator and rotor have two poles. Thus, the electrical system is rotating at ω_e and it thinks the rotor speed is ω_r since we assumed it to be a two-pole stator and rotor. Therefore, ω_r is now the rotor speed "referred" to the angular velocity of the

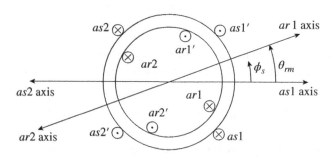

Figure 3.6-2 Mutual coupling between four-pole stator, *as*, and rotor, *ar*, windings.

electrical system and (3.6-8) applies regardless of how many even-number poles are on the stator and rotor.

Although we see that if we substitute (3.6-8) into (3.6-7), the result is the mutual inductance for a two-pole system as given by (3.4-32); however, what about the mutual inductance between stator windings as shown in Fig. 3.6-1? In the case of the two-pole machine, $L_{asbs} = 0$, since the magnetic axes are orthogonal. Note, however, that in the four-pole machine in Fig. 3.6-1, the bs winding is displaced 45° counterclockwise from the as winding. If we assume, for a minute, that in Fig. 3.6-1, we rotated the bs-winding clockwise 45°, it would be atop the as-winding and L_{asbs} would be similar to (3.6-7) except L_{asar} would be replaced by L_{asbs}, and L_{sr} would be replaced by L_{ms} and θ_{rm} by the angle between as and bs windings. If now we rotate the bs-winding counterclockwise by an angle of 45°, then L_{asbs} would be zero. From here on out, we will treat all electric machines as two-pole devices, knowing full well that (3.5-8) is the relationship between θ_r and θ_{rm} and

$$\frac{P}{2}\omega_{rm} = \omega_r \qquad (3.6\text{-}9)$$

where ω_r is referred to as the electrical angular velocity of the rotor.

Now, since we are going to treat all machines as two-pole devices and use θ_r in all the voltage equations, we have to do one last thing in order to make things work out. In Section 3.5, we expressed the change in energy entering the coupling field from the mechanical system as

$$dW_m = -T_e d\theta_r \qquad (3.6\text{-}10)$$

and the torque T_e as

$$T_e(\mathbf{i}, \theta_r) = \frac{\partial W_c(\mathbf{i}, \theta_r)}{\partial \theta_r} \qquad (3.6\text{-}11)$$

Equation (3.6-10) must now be written in terms of θ_{rm},

$$dW_m = -T_e d\theta_{rm} \qquad (3.6\text{-}12)$$

and since $\theta_{rm} = \frac{2}{P}\theta_r$, (3.5-11) becomes

$$T_e(\mathbf{i}, \theta_r) = \frac{P}{2} \frac{\partial W_c(\mathbf{i}, \theta_r)}{\partial \theta_r} \qquad (3.6\text{-}13)$$

Therefore, all expressions of torque, T_e, must be multiplied by $\frac{P}{2}$.

The torque and speed may be related by

$$T_e = J\left(\frac{2}{P}\right)\frac{d\omega_r}{dt} + B_m\left(\frac{2}{P}\right)\omega_r + T_L \qquad (3.6\text{-}14)$$

where P is the number of poles, J is the inertia in kg·m^2, B_m is the damping coefficient in N · m · s/rad, and T_L is positive for a load torque.

Example 3.B Torque Calculation for Four-Pole Machine

Consider the four-pole device shown in Fig. 3.6-2. $L_{sr} = 0.1$ H, $N_s = 2N_r$, $L_{ls} = 2L_{lr}$ where $L_{ls} = 0.1L_{ms}$, $i_{as} = 2i_{ar} = 2$ A. Determine the coenergy W_c and the torque T_e. With $\mathcal{R}_m$ the total reluctance of the two air gaps.

$$L_{sr} = \frac{N_s N_r}{\mathcal{R}_m} = 0.1 \text{ H} \tag{3B-1}$$

$$L_{ms} = \frac{N_s^2}{\mathcal{R}_m} \tag{3B-2}$$

$$L_{mr} = \frac{N_r^2}{\mathcal{R}_m} \tag{3B-3}$$

Therefore,

$$L_{ms} = \frac{N_s}{N_r} L_{sr} \tag{3B-4}$$
$$= 2L_{sr} = 0.2 \text{ H}$$

and

$$L_{mr} = L_{ms} \left(\frac{N_r}{N_s} \right)^2 \tag{3B-5}$$
$$= L_{ms} \left(\frac{1}{2} \right)^2 = 0.05 \text{ H}$$

Also,

$$L_{ls} = 0.1 \, L_{ms} = 0.02 \text{ H} \tag{3B-6}$$
$$L_{lr} = 0.5 \, L_{ls} = 0.01 \text{ H} \tag{3B-7}$$

Now,

$$W_c = \frac{1}{2} L_{ss} i_{as}^2 + L_{asar} i_{as} i_{bs} + \frac{1}{2} L_{rr} i_{br}^2 \tag{3B-8}$$

where

$$L_{ss} = L_{ls} + L_{ms} = 0.02 + 0.2 = 0.22 \text{ H} \tag{3B-9}$$
$$L_{asar} = 0.1 \cos \theta_r \tag{3B-10}$$
$$L_{rr} = L_{lr} + L_{mr} = 0.01 + 0.05 = 0.06 \text{ H} \tag{3B-11}$$

$$W_c = \left(\frac{1}{2}\right)(0.22)(2)^2 + (0.1)(2)(1)\cos\theta_r + \left(\frac{1}{2}\right)(0.06)(1)^2 \tag{3B-12}$$
$$= 0.44 + 0.2\cos\theta_r + 0.03 = 0.47 + 0.2\cos\theta_r \, \text{J}$$

The torque may be calculated using (3.6-13), that is,

$$T_e(\mathbf{i}, \theta_r) = \frac{P}{2}\frac{\partial W_c(\mathbf{i}, \theta_r)}{\partial\theta_r}$$
$$= \left(\frac{4}{2}\right)\frac{\partial(0.2\cos\theta_r)}{\partial\theta_r} \tag{3B-13}$$
$$= -0.4\sin\theta_r \, \text{N·m}$$

or

$$T_e(\mathbf{i}, \theta_{rm}) = -0.4\sin\frac{P}{2}\theta_{rm} \tag{3B-14}$$

SP3.6-1. A six-pole, 60 Hz induction motor is operating with rated frequency applied voltages. What is the speed in rad/s of mmf$_s$? [125.7 rad/s]

SP3.6-2. Suppose $\theta_{esi}(0) = 45°$ in (3.6-3) and (3.6-4). For a six-pole two-phase stator, determine the location of the positive and negative maximum values of mmf$_s^s$ at time zero. [S^s at $\phi_s = 15°,135°,255°$; N^s at $\phi_s = 75°,195°,315°$]

SP3.6-3. The rotor speed ω_{rm} of a six-pole 60 Hz two-phase induction motor is $0.3\omega_e$. Express (a) I'_{ar} (b) I''^s_{qr}, and (c) $\tilde{I}'_{ar}$ for balanced steady-state operation with $\tilde{I}'^s_{qr} = I'_r\underline{/30°}$. [(a) $I'_{ar} = \sqrt{2}I'_r\cos(0.0333\omega_et + 30°)$; (b) $I''^s_{qr} = \sqrt{2}I'_r\cos(\omega_et + 30°)$; (c) $\tilde{I}'_{ar} = \tilde{I}'^s_{qr}$]

3.7 Free Acceleration Variables Viewed from Different Reference Frames

It is instructive to view the free acceleration variables from the stationary, rotor, and synchronous reference frames. Free acceleration is when full voltage is applied at stall with $T_L = 0$ and without rotational losses represented. We will see that the torque appears the same in all reference frames and the voltage and currents form a balance set in all asynchronously rotating reference frames with a steady-state frequency equal to $\omega_e - \omega$, while in the synchronous reference frame the steady-state variables are dc. The plots shown in Figs. 3.7-1 and 3.7-2 are the actual machine variables during free acceleration. The torque versus rotor speed is shown in Fig. 3.7-2. The performance characteristics given in this section are for the single-fed two-pole two-phase 5-hp 110-V (rms) 60-Hz induction

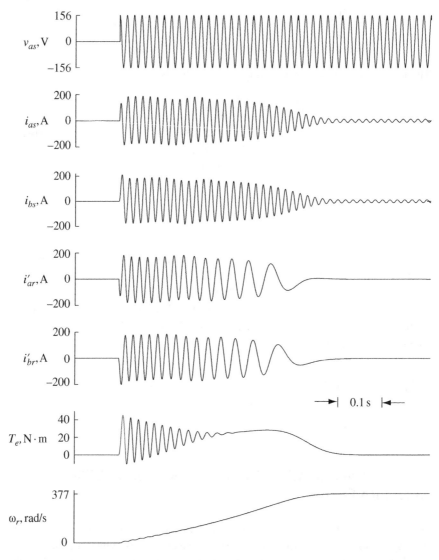

Figure 3.7-1 Free-acceleration characteristics of a two-pole two-phase 5-hp induction motor – actual machine variables.

machine with the following parameters: $r_s = 0.295\,\Omega$, $L_{ls} = 0.944\,\text{mH}$, $L_{ms} = 35.15\,\text{mH}$, $r'_r = 0.201\,\Omega$, and $L'_{lr} = 0.944\,\text{mH}$. The inertia of the rotor and connected mechanical load is $J = 0.026\,\text{kg·m}^2$.

The plots shown in Figs. 3.7-3 through 3.7-5 are for the same free acceleration as viewed from the stationary, synchronous, and rotor reference frames. Torque is the

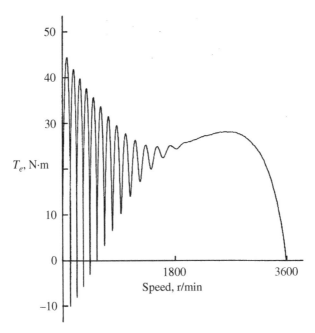

Figure 3.7-2 Torque versus speed during free-acceleration shown in Fig. 3.7-1.

same regardless of the reference frame from which it is being viewed; and all steady-state voltages and currents are varying at the same frequency in each reference frame. This is necessary since the change of variables provides a portrayal of the rotating mmfs from that reference frame. That is, mmf_s rotates at ω_e and is established by balanced currents flowing in the as- and bs-windings of frequency ω_e. The fictitious circuits for i_{qs}^s and i_{ds}^s are also fixed in the stator; the stationary reference frame. Therefore, these currents would vary at ω_e the same as i_{as} and i_{bs}. The fictitious windings for the rotor currents i_{qr}^s and i_{dr}^s are also flowing in stationary fictitious circuits; therefore, they would be balanced and vary at ω_e in order for mmf_r to rotate at ω_e. Another way to look at this is the actual rotor currents vary at $\omega_e - \omega_r$. When viewed from the stationary reference frame where the fictitious rotor windings are located, the frequency of the substitutive variables will be $\omega_e - \omega_r + \omega_r = \omega_e$ in order to produce a rotating magnetic field rotating at ω_e. The free acceleration as viewed from the synchronous reference frame is shown in Fig. 3.7-4. In this reference frame, the fictitious circuits are rotating at ω_e; therefore, since both mmf_s^e and mmf_r^e are constant in the steady state, the currents flowing in these fictitious circuits must be constant in the steady state for a given rotor

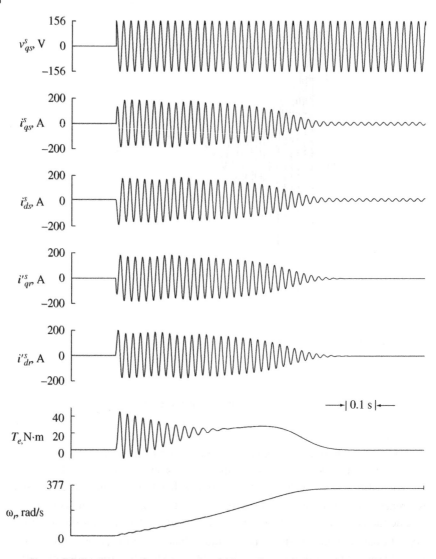

Figure 3.7-3 Free acceleration as viewed from the stationary reference frame.

speed. The transient offsets in the stator and rotor currents appear as small ω_e variations in the synchronously rotating reference frame currents.

The free acceleration as viewed from the rotor reference frame is shown in Fig. 3.7-5. This is an interesting reference frame since its speed varies as the rotor

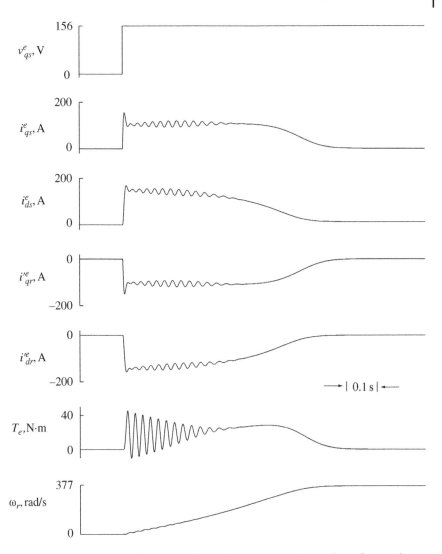

Figure 3.7-4 Free acceleration viewed from the synchronously rotating reference frame.

accelerates from $\omega_r = 0$ (stationary reference frame) to $\omega_r = \omega_e$ (synchronously rotating reference frame). Therefore, the fictitious circuits are stationary and the currents start with a frequency of ω_e and i_{qs}^r and i_{ds}^r end as constants and i_{qr}^r and i_{dr}^r end as zeros. Since mmf$_r^e$ is zero at $\omega_r = \omega_e$.

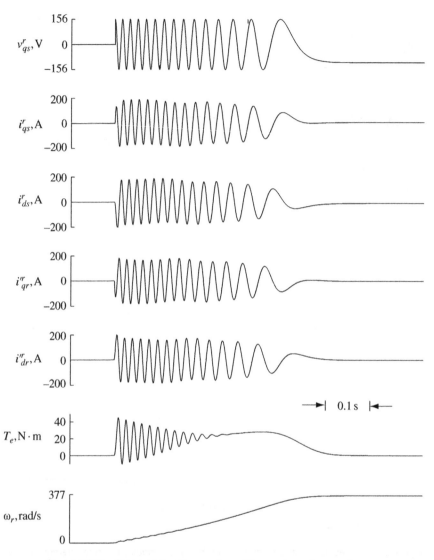

Figure 3.7-5 Free acceleration as viewed from rotor reference frame.

Example 3.C   Write the Voltage Equations in the Synchronously Rotating Reference Frame

From Fig. 3.4-2 with $\omega = \omega_e$.

$$v_{qs}^e = r_s i_{qs}^e + \omega_e \left(L_{ss} i_{ds}^e + L_{ms} i_{dr}'^e \right) + L_{ss} \frac{di_{qs}^e}{dt} + L_{ms} \frac{di_{qr}'^e}{dt} \qquad (3\text{C-1})$$

$$v_{ds}^e = r_s i_{ds}^e - \omega_e \left(L_{ss} i_{qs}^e + L_{ms} i_{qr}^{\prime e} \right) + L_{ss} \frac{di_{ds}^e}{dt} + L_{ms} \frac{di_{dr}^{\prime e}}{dt} \tag{3C-2}$$

$$v_{qr}^{\prime e} = r_r^\prime i_{qr}^{\prime e} + (\omega_e - \omega_r)\left(L_{rr}^\prime i_{dr}^{\prime e} + L_{ms} i_{ds}^e \right) + L_{rr}^\prime \frac{di_{qr}^{\prime e}}{dt} + L_{ms} \frac{di_{qs}^e}{dt} \tag{3C-3}$$

$$v_{dr}^{\prime e} = r_r^\prime i_{dr}^{\prime e} - (\omega_e - \omega_r)\left(L_{rr}^\prime i_{qr}^{\prime e} + L_{ms} i_{qs}^e \right) + L_{rr}^\prime \frac{di_{dr}^{\prime e}}{dt} + L_{ms} \frac{di_{ds}^e}{dt} \tag{3C-4}$$

SP3.7-1. Determine the steady-state frequency for a given speed of (a) $i_{ar}^\prime$ and $i_{br}^\prime$ in Fig. 3.7-1, (b) $i_{qr}^{\prime s}$ and $i_{dr}^{\prime s}$ in Fig. 3.7-3, and (c) i_{qr}^e and i_{dr}^e in Fig. 3.7-4. [(a) $\omega_e - \omega_r$, (b) ω_e and zero at $\omega_r = \omega_e$, (c) dc for $\omega_r \neq \omega_e$ and zero for $\omega_r = \omega_e$]

SP3.7-2. Repeat SP3.7-1 for the stator-related variables. [(a) ω_e, (b) ω_e, (c) dc]

SP3.7-3. Why are v_{qs}^e in Figs. 3.7-4 and 3.7-5 different? [In Fig. 3.7-4 $\omega = \omega_e$ from $\omega_r = 0$ in Fig. 3.7-5 ω_r becomes ω_e at $\omega_r = \omega_e$ and v_{qs}^e becomes the value corresponding to values at time ω_r became equal to ω_e]

3.8 Steady-State Equivalent Circuit

In this section, we will set forth the instantaneous and steady-state phasors and the steady-state equivalent circuit. The instantaneous phasor equations may be obtained by a procedure used in Chapter 2. In particular, we will substitute the voltage equations in the synchronously rotating reference frame given by (3C-1) through (3C-4) into

$$\tilde{f}_{as} = f_{qs}^e - jf_{ds}^e \tag{3.8-1}$$

$$\tilde{f}_{ar}^\prime = f_{qr}^{\prime e} - jf_{dr}^{\prime e} \tag{3.8-2}$$

The instantaneous phasor voltage equations from Example 3.C which are in the synchronous reference frame become

$$\tilde{v}_{as} = r_s \tilde{i}_{as} + j\omega_e L_{ss} \tilde{i}_{as} + j\omega_e L_{ms} \tilde{i}_{ar}^\prime + pL_{ss} \tilde{i}_{as} + pL_{ms} \tilde{i}_{ar}^\prime \tag{3.8-3}$$

$$\tilde{v}_{ar}^\prime = r_s^\prime \tilde{i}_{ar}^\prime + j(\omega_e - \omega_r)L_{rr}^\prime \tilde{i}_{ar}^\prime + j(\omega_e - \omega_r)L_{ms} \tilde{i}_{as} \\ + pL_{rr}^\prime \tilde{i}_{ar}^\prime + pL_{ms} \tilde{i}_{as} \tag{3.8-4}$$

For steady-state conditions the last two terms of (3.8-3) and (3.8-4) become zero and we obtain the following steady-state voltage equations which are rotating at ω_e as a phasor

$$\tilde{V}_{as} = r_s \tilde{I}_{as} + j\omega_e (L_{ls} + L_{ms})\tilde{I}_{as} + j\omega_e L_{ms} \tilde{I}_{ar}^\prime \tag{3.8-5}$$

$$\tilde{V}'_{ar} = r'_r\tilde{I}'_{ar} + j(\omega_e - \omega_r)(L'_{lr} + L_{ms})\tilde{I}'_{ar} + j(\omega_e - \omega_r)L_{ms}\tilde{I}_{as} \qquad (3.8\text{-}6)$$

The so-called "slip" is

$$s = \frac{\omega_e - \omega_r}{\omega_e} \qquad (3.8\text{-}7)$$

We see that slip increases when load torque increases, also, if we divide (3.8-6) by the slip, it becomes

$$\frac{\tilde{V}'_{ar}}{s} = \frac{r'_r}{s}\tilde{I}'_{ar} + j\omega_e(L'_{lr} + L_{ms})\tilde{I}'_{ar} + j\omega_eL_{ms}\tilde{I}_{as} \qquad (3.8\text{-}8)$$

Equations (3.8-5) and (3.8-8) suggest the single-phase equivalent T circuit of a two-phase symmetrical machine during steady-state balanced operation shown in Fig. 3.8-1. Please note that the inductive reactances are calculated as $X = \omega_e L$. One tends to want to calculate the inductive reactances of the rotor circuit as $X = (\omega_e - \omega_r)L$ and (3.8-6) is of the form we would expect; however, we have divided (3.8-6) by $(\omega_e - \omega_r)$ and multiplied by ω_e to arrive at (3.8-8). With $\tilde{V}'_{ar}$ equal to zero, only $\frac{r'_r}{s}$ changes with rotor speed.

We understand that current is not induced in the rotor windings when $\omega_r = \omega_e$. Since the rotor windings are generally short-circuited (v'_{ar} and $v'_{br} = 0$) and from (3.8-7) the slip is zero and $\frac{r'_r}{s}$ is infinite; hence, the rotor circuit appears to be open-circuited, thus correctly portraying synchronous speed "operation." We will find that, with a slight modification (X_{ms} becomes $\frac{3}{2}X_{ms}$), this circuit may also be used to calculate the per-phase steady-state performance of a three-phase symmetrical machine.

An expression for the steady-state electromagnetic torque may be obtained by first writing (3.5-19) in terms of I^e_{qs}, I^e_{ds}, I'^e_{qr}, and I'^e_{dr}, and then express $\tilde{I}_{as}$ and $\tilde{I}'_{ar}$ using

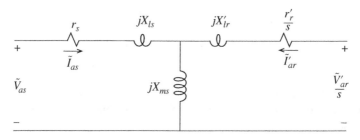

Figure 3.8-1 Equivalent single-phase circuit for a two-phase symmetrical induction machine for balanced steady-state operation.

$$\sqrt{2}\tilde{F}_{as} = F^e_{qs} - jF^e_{ds} \tag{3.8-9}$$

$$\sqrt{2}\tilde{F}_{ar} = F^e_{qr} - jF^e_{dr} \tag{3.8-10}$$

The expression may be reduced to

$$T_e = N_p \left(\frac{P}{2}\right) L_{ms} \operatorname{Re}\left[j\tilde{I}^*_{as}\tilde{I}'_{ar}\right] \tag{3.8-11}$$

where $\tilde{I}^*_{as}$ is the conjugate of $\tilde{I}_{as}$. The phasor currents may be calculated from the equivalent circuit given in Fig. 3.8-1. For a three-phase machine L_{ms} must be multiplied by $\frac{3}{2}$ and N_p is either 2 or 3 depending on the number of phases.

The balanced steady-state torque-speed or torque-slip characteristic of a single-fed induction machine warrants discussion. The vast majority of induction machines in use today are single-fed, wherein electric power is transferred to or from the induction machine via the stator circuits with the rotor windings short-circuited. Therefore, $\tilde{V}'_{ar}$ is zero, whereupon (3.8-8) may be written as

$$\tilde{I}'_{ar} = -\frac{jX_{ms}}{r'_r/s + j(X'_{lr} + X_{ms})}\tilde{I}_{as} \tag{3.8-12}$$

Substituting (3.8-12) into (3.8-11) yields the following expression for electromagnetic torque of a single-fed two-phase symmetrical induction machine during balanced steady-state operation:

$$T_e = \frac{N_p(P/2)\left(X^2_{ms}/\omega_e\right)\left(r'_r/s\right)|\tilde{I}_{as}|^2}{\left(r'_r/s\right)^2 + \left(X'_{lr} + X_{ms}\right)^2} \tag{3.8-13}$$

In (3.8-12) and (3.8-13), X_{ms} is replaced by $\frac{3}{2}X_{ms}$ for a three-phase machine.

It is important to note from (3.8-13) that torque is positive (motor action) when slip is positive which occurs when $\omega_r < \omega_e$, negative (generator action) when the slip is negative which occurs when the rotor is being driven above synchronous speed, $\omega_r > \omega_e$, and zero when the slip is zero ($\omega_r = \omega_e$). In other words, the single-fed induction machine develops torque at all speeds except at synchronous speed.

With the rotor windings short-circuited, the input impedance of the equivalent circuit shown in Fig. 3.8-1 is

$$Z = \frac{\left(r_s r'_r/s\right) + \left(X^2_{ms} - X_{ss}X'_{rr}\right) + j\left[\left(r'_r/s\right)X_{ss} + r_sX'_{rr}\right]}{\left(r'_r/s\right) + jX'_{rr}} \tag{3.8-14}$$

For a three-phase machine, X_{ms} is replaced with $\frac{3}{2}X_{ms}$. Now $|\tilde{I}_{as}|^2$ is I^2_s and

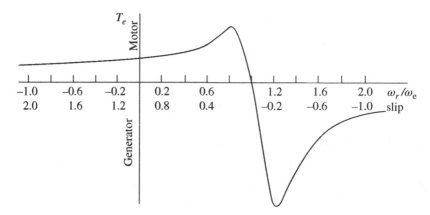

Figure 3.8-2 Steady-state torque-speed characteristics of a symmetrical induction machine.

$$I_s = \frac{|\tilde{V}_{as}|}{|Z|} \tag{3.8-15}$$

Hence, the expression for the steady-state electromagnetic torque for a single-fed two-phase symmetrical induction machine becomes

$$T_e = \frac{N_p(P/2)\left(X_{ms}^2/\omega_e\right)r'_r s|\tilde{V}_{as}|^2}{\left[r_s r'_r + s\left(X_{ms}^2 - X_{ss}X'_{rr}\right)\right]^2 + \left(r'_r X_{ss} + s r_s X'_{rr}\right)^2} \tag{3.8-16}$$

Again, for a three-phase machine, X_{ms} is replaced with $\frac{3}{2}X_{ms}$. Thus, for a given set of parameters and source frequency ω_e, the steady-state torque varies as the square of the magnitude of the applied voltages.

Figure 3.8-2 shows the steady-state torque speed plot of a typical industrial-type induction motor. Stable operation occurs on the negative slope part of this plot.

Example 3.D   Phasor Diagram for Steady-State Operation

Consider a single-fed two-pole two-phase 5-hp 110-V (rms) 60-Hz induction machine with the following parameters: $r_s = 0.295\,\Omega$, $L_{ls} = 0.944\,\text{mH}$, $L_{ms} = 35.15\,\text{mH}$, $r'_r = 0.201\,\Omega$, and $L'_{lr} = 0.944\,\text{mH}$. Calculate $\tilde{I}_{as}$, $\tilde{I}_{ar}$, T_e, draw the phasor diagram, and show the rotor and stator poles for (a) $s = 0.05$ and (b) $s = -0.05$. The equivalent circuit for (a) and (b) is shown in Fig. 3.D-1. The inductive reactances are calculated as ω_e times the inductance in henrys.

$$r_s = 0.295\,\Omega \quad X_{ms} = 13.252\,\Omega \quad r'_r = 0.201\,\Omega$$
$$X_{ls} = 0.356\,\Omega \qquad\qquad\qquad\quad X'_{lr} = 0.356\,\Omega$$

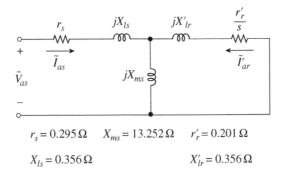

$r_s = 0.295\,\Omega$ $X_{ms} = 13.252\,\Omega$ $r'_r = 0.201\,\Omega$

$X_{ls} = 0.356\,\Omega$ $X'_{lr} = 0.356\,\Omega$

Figure 3.D-1 Equivalent circuit for steady-state operation of a single-fed induction machine.

For part (a):

$$\frac{r'_r}{s} = \frac{0.201}{0.05} = 4.02\,\Omega \tag{3D-1}$$

$$Z_{in} = r_s + jX_{ls} + \frac{\left(\dfrac{r'_r}{s} + jX'_{lr}\right)jX_{ms}}{\dfrac{r'_r}{s} + j\left(X'_{lr} + X_{ms}\right)} \tag{3D-2}$$

$$= 0.295 + j0.356 + \frac{(4.02 + j0.356)j13.252}{4.02 + j(0.356 + 13.252)} = 4.175\underline{/24.64°}\,\Omega$$

$$\tilde{I}_{as} = \frac{\tilde{V}_{as}}{Z_{in}} \tag{3D-3}$$

$$= \frac{110\underline{/0°}}{4.175\underline{/24.64°}} = 26.35\underline{/-24.64°}\ \text{A}$$

With $\tilde{V}_{ar} = 0$, $\tilde{I}_{ar}$ may be obtained from $\tilde{I}_{as}$ by current division. That is,

$$\tilde{I}'_{ar} = -\frac{jX_{ms}}{(r'_r/s) + jX'_{rr}}\tilde{I}_{as}$$

$$= -\frac{j13.252}{4.02 + j(0.356 + 13.252)}\,26.35\underline{/-24.64°} \tag{3D-4}$$

$$= -24.61\underline{/-8.18°} = 24.61\underline{/171.8°}\ \text{A}$$

From (3.8-11),

$$T_e = N_p \left(\frac{P}{2}\right) L_{ms} R_e \left[j\tilde{I}_{as}^* \tilde{I}_{ar}'\right]$$
$$= 2\left(\frac{2}{2}\right)(35.15 \times 10^{-3}) R_e \left[26.35\underline{/24.64} + 90 \ 24.61\underline{/171.8}\right] \tag{3D-5}$$
$$= 12.9 \, \text{N·m}$$

The torque is positive in the direction of rotation, motor action. The phasor diagram for part (a) is shown in Fig. 3.D-2. The stator poles are "pushing" the rotor poles counterclockwise, motor action.

For part (b):

$$\frac{r_r'}{-0.05} = -4.02 \, \Omega \tag{3D-6}$$

In the calculation, we need to replace $\dfrac{r_r'}{s} = 0.402 \, \Omega$ in part (a) with $\dfrac{r_r'}{s} = -0.402 \, \Omega$. Thus,

$$Z_{in} = 3.648\underline{/151.5°} \, \Omega \tag{3D-7}$$

$$\tilde{I}_{as} = 30.154\underline{/-151.5°} \, \text{A} \tag{3D-8}$$

$$\tilde{I}_{ar}' = 28.2\underline{/12°} \, \text{A} \tag{3D-9}$$

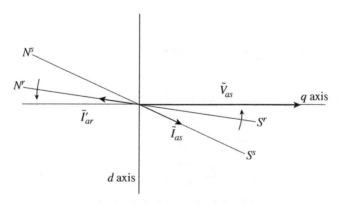

Figure 3.D-2 Phasor diagram, motor action.

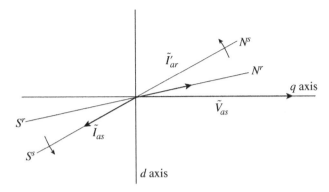

Figure 3.D-3 Phasor diagram, generator action.

From (3.8-11),

$$T_e = 2\left(\frac{2}{2}\right)(35.15 \times 10^{-3})R_e\left[j30.15\underline{/151.5}\,28.2\underline{/12}\right] \tag{3D-10}$$
$$= -17.0\,\text{N·m}$$

The torque is negative in the direction of rotation, generator action. The phasor diagram is shown in Fig. 3.D-3. Therein, the rotor poles are "pushing" the stator poles counterclockwise, generator action.

We can determine motor or generator action using the cross product rule (3.5-26). Therein, A is $\dfrac{\text{mmf}_s}{\mathcal{R}_m}$ and B is mmf_r. From Fig. 3.D-2, AxB is a torque vector positive for motor action out of the paper. In Fig. 3.D-3, AxB is a torque vector in the opposite direction, into the paper, generator action.

In most cases, the load torque is a function of ω_r, say $T_L = K\omega_r^2$, for example. In these cases, the machine can develop sufficient starting torque and, if T_L and T_e match on the negative slope portion, stable operation will occur. If, on the other hand, T_L is constant and greater than T_e at $\omega_r = 0$, we have at least three choices: (1) increase the stator voltage; (2) increase the rotor resistance; or (3) use a different machine. Increasing the rotor resistance to increase the starting torque is something that we have not yet discussed. We will now.

An expression for the slip at maximum torque may be obtained by taking the derivative of (3.8-16) with respect to slip and setting the result equal to zero. In particular,

$$s_m = r_r'G \tag{3.8-17}$$

where s_m is the slip at maximum torque and

$$G = \pm\sqrt{\frac{r_s^2 + X_{ss}^2}{\left(X_{ms}^2 - X_{ss}X_{rr}'\right)^2 + r_s^2 X_{rr}'^2}} \tag{3.8-18}$$

where X_{ms} is replaced by $\dfrac{3}{2}X_{ms}$ for a three-phase machine. Two values of slip at maximum torque, s_m, are obtained, one for motor action and one for generator action. It is important to note that G is not a function of r'_r; thus, the slip at maximum torque, (3.8-17), is directly proportional to r'_r. Consequently, with all other machine parameters constant, the speed at which maximum steady-state torque occurs may be varied by inserting external rotor resistance. This feature is often used when starting large motors which have coil-wound rotor windings with slip rings. In this application, balanced external rotor resistances are placed across the terminals of the rotor windings so that maximum torque occurs near stall. As the machine speeds up, the external resistors are decreased in value. On the other hand, some induction machines are designed with high-resistance rotor windings so that maximum torque is produced at or near stall to provide fast response.

It may at first appear that the magnitude of the maximum torque would be influenced by r'_r. However, if (3.7-12) is substituted into (3.7-16), the maximum torque may be expressed as

$$T_{e,\max} = \frac{N_p(P/2)\left(X^2_{ms}/\omega_e\right)G\left|\tilde{V}_{as}\right|^2}{\left[r_s + G\left(X^2_{ms} - X_{ss}X'_{rr}\right)\right]^2 + \left(X_{ss} + Gr_sX'_{rr}\right)^2} \tag{3.8-19}$$

where X_{ms} is replaced by $\frac{3}{2}X_{ms}$ for a three-phase machine and N_p is the number of phases, either two or three. Equation (3.8-19) is independent of r'_r. Thus, the maximum torque remains constant if only r'_r is varied; however, the slip at which maximum torque is produced varies in accordance with (3.8-17). Figure 3.8-3 illustrates the effect of changing r'_r. Therein, $r'_{r3} > r'_{r2} > r'_{r1}$.

In variable-frequency drive systems, the operating speed of the induction machine is controlled by changing the frequency of the applied voltages by either a converter (solid-state dc-to-ac converter) or a cycloconverter (ac frequency changer). The phasor voltage equations are applicable regardless of the frequency of operation. It is only necessary to keep in mind that the reactances given in the steady-state equivalent circuit, Fig. 3.8-1, are defined as the product of ω_e and the inductances. As the frequency is decreased, the time rate of change of the steady-state variables is decreased proportionally. Thus, the inductive reactances decrease linearly with frequency. If the amplitude of the applied voltages is maintained at the rated value, the currents will become excessive at the lower frequencies. To prevent these large currents, the magnitude of the stator voltages is decreased as the frequency is decreased. In many applications, the voltage magnitude is reduced linearly with frequency until a low frequency is reached, whereupon the decrease in voltage is programmed in a manner to compensate for the effects of the stator resistance.

The influence of frequency upon the steady-state torque-speed characteristics is illustrated in Fig. 3.8-4. These characteristics are for a linear relationship between

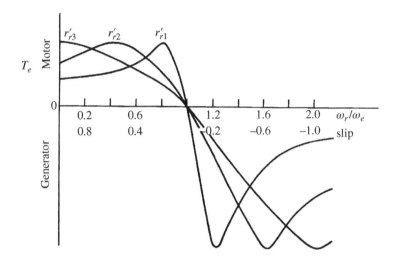

Figure 3.8-3 Steady-state torque-speed characteristics of a symmetrical induction machine for different values of r_r'.

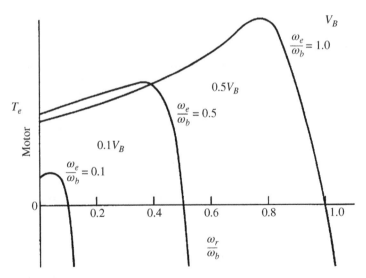

Figure 3.8-4 Steady-state torque-speed characteristics of a symmetrical induction machine for different operating frequencies.

the magnitude of the applied voltages and frequency. This machine is designed to operate at $\omega_e = \omega_b$, where ω_b corresponds to the rated frequency. Rated voltage is applied at rated frequency, that is, when $\omega_e = \omega_b$, $|\tilde{V}_{as}| = V_B$, where V_B is the base or rated voltage. Since the reactances ($\omega_e L$) decrease with frequency, the voltage is reduced as frequency is reduced to avoid large stator currents. The maximum torque is reduced markedly at $\omega_e / \omega_b = 0.1$. At this frequency, the voltage would probably be increased somewhat so as to obtain a higher torque. Perhaps a voltage of, say, $0.15 V_B$ or $0.2 V_B$ would be used rather than $0.1 V_B$. Saturation of the stator or rotor steel may, however, cause the stator currents to be excessive at this higher voltage. These practical considerations of variable-frequency drives are of major importance but beyond the scope of this text. However, we will encounter variable frequency operation later when we deal with field orientation of an induction machine.

Example 3.E   No-Load and Blocked-Rotor Tests

The parameters for the equivalent circuit shown in Fig. 3.8-1 may be calculated by using electromagnetic field theory or determined from tests. The tests generally performed are a dc test, a no-load test, and a blocked-rotor test. Table 3.E-1 gives the test data for a 3-hp four-pole 110-V (rms) two-phase 60-Hz induction machine, where all ac voltages and currents are in rms values:

Table 3.E-1   Test data to determine machine parameters.

DC test	No-load test	Blocked-rotor test
$V_{dc} = 6.9\,\text{V}$	$V_{nl} = 110\,\text{V}$	$V_{br} = 23.5\,\text{V}$
$I_{dc} = 13.0\,\text{A}$	$I_{nl} = 3.86\,\text{A}$	$I_{br} = 16.1\,\text{A}$
	$P_{nl} = 134\,\text{W}$	$P_{br} = 469\,\text{W}$
	$f = 60\,\text{Hz}$	$f = 15\,\text{Hz}$

During the dc test, a dc voltage is applied to one phase while the machine is at standstill. Thus,

$$r_s = \frac{V_{dc}}{I_{dc}} = \frac{6.9}{13} = 0.531\,\Omega \tag{3E-1}$$

The no-load test, which is analogous to the transformer open-circuit test, is performed with balanced 60-Hz voltages applied to the stator windings without mechanical load on the rotor (no load). The total power input during this test is the sum of the stator ohmic losses per phase, the core losses due to hysteresis and eddy currents, and rotational losses due to friction and windage. The total stator ohmic losses (two phases) are

$$P_r = N_p I_{nl}^2 r_s = 2(3.86)^2(0.531) = 15.8 \text{ W} \tag{3E-2}$$

Therefore, the power loss due to friction, windage, and core losses is

$$P_{fWC} = P_{nl} - P_r = 134 - 15.8 = 118.2 \text{ W} \tag{3E-3}$$

In the equivalent circuit shown in Fig. 3.8-1, the core loss is neglected. It is generally small and, in most cases, little error is introduced by neglecting it. It can be approximated by placing a resistor in shunt with the magnetizing reactance X_{ms}. The friction and windage losses may be approximated with B_m in (3.5-8).

It is noted from the no-load test data that the power factor is very small since the total apparent two-phase power input, S_{nl}, to the motor is

$$|S_{nl}| = N_p V_{nl} I_{nl} = 2(110)(3.86) = 849.2 \text{ VA} \tag{3E-4}$$

Therefore, the no-load impedance is highly inductive, and its magnitude is assumed to be the sum of the stator leakage reactance and the magnetizing reactance since the rotor speed is essentially synchronous, $(s \approx 0)$, whereupon r_r'/s is much larger than X_{ms} in Fig. 3.8-1. Thus,

$$X_{ls} + X_{ms} \cong \frac{V_{nl}}{I_{nl}} = \frac{110}{3.86} = 28.5 \,\Omega \tag{3E-5}$$

During the blocked-rotor test, which is analogous to the transformer short-circuit test, the rotor is locked by some mechanical means and balanced two-phase stator voltages are applied. The frequency of the applied voltages is often less than rated in order to obtain a representative value of r_r', since during normal operation the frequency of the rotor currents is low and the rotor resistance of some induction machines varies considerably with frequency. During stall $(s = 1)$, the rotor impedance $r_r'/s + jX_{lr}'$ is much smaller in magnitude than X_{ms}, whereupon the current flowing in the magnetizing reactance may be neglected in these calculations. Hence, the total two-phase power input to the motor during the blocked-rotor test is

$$P_{br} = N_p I_{br}^2 (r_s + r_r') \tag{3E-6}$$

From which

$$r'_r = \frac{P_{br}}{2I^2_{br}} - r_s = \frac{469}{(2)(16.1)^2} - 0.531 = 0.374\,\Omega \tag{3E-7}$$

The magnitude of the blocked-rotor input impedance is

$$|Z_{br}| = \frac{V_{br}}{I_{br}} = \frac{23.5}{16.1} = 1.46\,\Omega \tag{3E-8}$$

Thus, since the frequency of the applied voltage for the block-rotor test is 15 Hz,

$$\left|(r_s + r'_r) + j\frac{15}{60}(X_{ls} + X'_{lr})\right| = 1.46\,\Omega \tag{3E-9}$$

From which

$$\left[\frac{15}{60}(X_{ls} + X'_{lr})\right]^2 = (1.46)^2 - (r_s + r'_r)^2$$
$$= (1.46)^2 - (0.531 + 0.374)^2 = 1.31\,\Omega \tag{3E-10}$$

Thus,

$$X_{ls} + X'_{lr} = 4.58\,\Omega \tag{3E-11}$$

Generally, X_{ls} and X'_{lr} are assumed equal; however, in some types of induction machines, a different ratio is suggested. We will assume $X_{ls} = X'_{lr}$, whereupon we have determined the machine parameters. In particular, for $\omega_e = 377\,\text{rad/s}$, the parameters are $r_s = 0.531\,\Omega$, $X_{ls} = 2.29\,\Omega$, $X_{ms} = 26.2\,\Omega$, $r'_r = 0.374\,\Omega$, and $X'_{lr} = 2.29\,\Omega$.

Example 3.F   Starting Torque and Current Calculations

A four-pole 110-V (rms) 28-A 7.5-hp two-phase induction motor has the following parameters: $r_s = 0.3\,\Omega$, $L_{ls} = 0.0015\,\text{H}$, $L_{ms} = 0.035\,\text{H}$, $r'_r = 0.15\,\Omega$, and $L'_{lr} = 0.0007\,\text{H}$. The machine is supplied from a 110-V 60-Hz source. Calculate the starting torque and starting current.

It would be convenient to use a computer to solve for the starting current and torque if the electrical and mechanical transients are to be considered. However, an approximation of the actual starting characteristics may be obtained from a constant-speed steady-state analysis. For this purpose, it is assumed that the speed is fixed at zero and the electric system has established steady-state operation.

$$X_{ss} = \omega_e(L_{ls} + L_{ms}) = 377(0.0015 + 0.035) = 13.76\,\Omega \tag{3F-1}$$

$$X'_{rr} = \omega_e(L'_{lr} + L_{ms}) = 377(0.0007 + 0.035) = 13.46\,\Omega \tag{3F-2}$$

$$X_{ms} = \omega_e L_{ms} = 377(0.035) = 13.2\,\Omega \tag{3F-3}$$

The steady-state torque with $\omega_r = 0$ ($s = 1$) may be calculated from (3.8-16).

$$T_e = \frac{N_p(P/2)\left(X_{ms}^2/\omega_e\right)r_r's\left|\tilde{V}_{as}\right|^2}{\left[r_sr_r' + s\left(X_{ms}^2 - X_{ss}X_{rr}'\right)\right]^2 + \left(r_r'X_{ss} + sr_sX_{rr}'\right)^2}$$

$$= \frac{2\left(\dfrac{4}{2}\right)(13.2^2/377)(0.15)(1)(110)^2}{\{(0.3)(0.15) + (1)[13.2^2 - (13.76)(13.46)]\}^2 + [(0.15)(13.76) + (1)(0.3)(13.46)]^2}$$

$$= 21.4\,\text{N}\cdot\text{m}$$

(3F-4)

Since $s = 1$, the rotor impedance in parallel with X_{ms} is much smaller than X_{ms}. Thus, for this mode of operation, the input impedance is approximately

$$\begin{aligned}
Z &= (r_s + r_r') + j\left(X_{ls} + X_{lr}'\right)\\
&= (0.3 + 0.15) + j377(0.0015 + 0.0007)\\
&= 0.45 + j0.83\,\Omega = 0.944\,\underline{/61.5°}\,\Omega
\end{aligned}$$

(3F-5)

With $\tilde{V}_{as}$ as the reference phasor, then

$$\tilde{I}_{as} = \frac{\tilde{V}_{as}}{Z} = \frac{110\underline{/0°}}{0.944\underline{/61.5°}} = 117\underline{/-61.5°}\,\text{A}$$

(3F-6)

The stall or starting current is over four times larger than the rated current. In some large machines, the starting current with rated voltage applied may be ten times the rated current. This high value of current may cause overheating and damage to the windings. Consequently, reduced voltage is applied to many large machines during the starting period, and rated voltage is not applied until the machine has accelerated to near rated speed.

SP3.8-1. Neglecting the current flowing in X_{ms} is generally an acceptable approximation when calculating the machine parameters from the blocked-rotor test (Example 3.E). However, this approximation is not valid when calculating the blocked-rotor torque. Use (3.8-11) and let $\tilde{I}_{as} = a + jb$ to show that T_e is zero regardless of the rotor speed if the current flowing in X_{ms} is assumed to be negligibly small. [Real parts cancel]

SP3.8-2. Assume that the friction, windage, and core losses P_{fWC}, calculated in Example 3.E, are to be represented by $B_m\omega_{rm}$, with B_m selected so that a load equivalent to 118.2 W occurs at $\omega_r = 0.9\omega_e$. Determine B_m [$B_m = 4.11 \times 10^{-3}$ N·m·s/rad]

SP3.8-3. Using the parameters given in Example 3.D, determine $\tilde{I}_{as}$ for (a) $\omega_r = \omega_e$ and (b) $\omega_r = 0$. Let $\tilde{I}_{as} = -\tilde{I}_{ar}$. [(a) $\tilde{I}_{as} = 8.06\underline{/-88.8°}$ A, (b) $\tilde{I}_{as} = 11.5\underline{/-53.1°}$ A]

SP3.8-4. The frequency of the balanced stator currents of a two-phase induction machine is 60 Hz, mmf$_s$ rotates counterclockwise. The device is operating as a motor, and the rotor of the two-pole machine is rotating counterclockwise at $0.9\omega_e$. (a) Determine the frequency of the balanced rotor currents. Determine the angular velocity of mmf$_s$ and mmf$_r$ relative to an observer sitting (b) on the rotor and (c) on the stator. [(a) 6 Hz; (b) 37.7 rad/s, ccw; (c) 377 rad/s, ccw]

SP3.8-5. Assume that θ_r is positive in the clockwise direction in Fig. 3.2-3 rather than in the counterclockwise direction. Express all inductances. [$L_{asbr} = L_{sr}\sin\theta_r$; $L_{bsar} = -L_{sr}\sin\theta_r$; all others unchanged]

3.9 Problems

1 Obtain (3.3-8) from (3.3-3).

2 Obtain (3.4-45).

3 Show that (3.5-31) is correct.

4 Show that (3.8-11) is correct.

5 Select a speed on the positive slope portion of T_e vs ω_r in Fig. 3.8-2 and show that this is an unstable point of operation, then select a speed on the negative slope and show that it is stable.

6 Consider the two-pole, two-phase induction machine shown in Fig. 3.2-3. The device is operating as a motor at $\omega_r = 95\pi$ rad/s with $I'_{ar} = \cos 5\pi t$ and $I'_{br} = -\sin 5\pi t$. Determine the angular velocity and direction of mmf$_r$ relative to (a) an observer on the rotor and (b) an observer on the stator. Also determine (c) angular velocity of the stator currents and (d) the direction of rotation of the rotor.

7 The windings shown in Fig. 3.9-1 are sinusoidally distributed and the device is symmetrical. The amplitude of the stator-to-rotor mutual inductance is L_{sr}. Express all mutual inductances as functions of L_{sr} and θ_r.

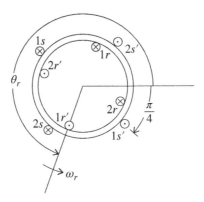

Figure 3.9-1 Coupled windings.

8 The rotor windings of the two-pole, two-phase symmetrical induction machine shown in Fig. 3.2-3 are open-circuited. $I_{as} = \sqrt{2}I_s \cos \omega_e t$ and $I_{bs} = -\sqrt{2}I_s \sin \omega_e t$. The rotor is driven at $\omega_r = \omega_e$ in the counterclockwise direction. Express V_{ar}.

9 Synchronously rotating variables may be related to rotor reference frame variables by $\mathbf{f}^e_{qds} = {}^r\mathbf{K}^e \mathbf{f}^r_{qds}$. Determine ${}^r\mathbf{K}^e$ in terms of $\mathbf{K}^e_s$ and $\mathbf{K}^r_s$.

10 A four-pole, two-phase induction machine has the following parameters: $r_s = 0.3\ \Omega$, $L_{ls} = 1\ \text{mH}$, $L_{ms} = 20\ \text{mH}$, $r'_r = 0.2\ \Omega$, and $L'_{lr} = 1\ \text{mH}$. The device is supplied from a 60-Hz source; the rotor speed is $\omega_r = 360\ \text{rad/s}$. In this mode of operation $\tilde{I}_{as} = 28.8\underline{/-36.1°}$ A and $\tilde{I}_{ar} = 23.9\underline{/-173.2°}$ A. Calculate (a) T_e, (b) the total ohmic loss in the rotor windings, (c) the mechanical power delivered to the load, and (d) express I_{as}, I'_{ar}, I^s_{qs}, and I'^s_{qr}.

11 Consider Problem 10. Calculate $\tilde{V}_{as}$. Draw the phasor diagram and locate the poles. Show all voltage drops associated with the equivalent circuit shown in Fig. 3.8-1.

12 Construct the mmf^e_s and mmf^e_r as shown in Fig. 3.3-2 for Figs. 3.D-2 and 3. D-3.

13 Select two identical capacitors so that when they are connected in parallel with each phase (one capacitor per phase) of the induction machine described in Example 3.E, the no-load capacitor-induction machine combination operates at unity power factor. Assume the capacitors are ideal (zero resistance).

14 Show from Fig. 3.4-2 that for balanced conditions that the rotor currents are zero when $\omega_r = \omega_e$.

References

1 P. C. Krause, *Analysis of Electric Machinery*, McGraw-Hill Book Company, New York, 1986.

2 D. C. White and H. H. Woodson, *Electromechanical Energy Conversion*, John Wiley and Sons, New York, 1959.

4

Synchronous Machines

4.1 Introduction

In this chapter, two types of synchronous machines are considered, the permanent magnet synchronous machine and the synchronous generator. The permanent magnet synchronous machine is the machine used in brushless dc drives where power electronics, specifically, an inverter, is used to change the frequency of the applied stator voltages such that they match the electrical angular velocity of the rotor. This type of drive is covered in Chapter 6 on electric drives. In this chapter, the voltage and flux linkage equations are derived for use in Chapter 6. If your interest is in electric drives, you need only to study Section 4.2 of this chapter. If your interest is in power systems, you need to study the complete chapter.

The stator and rotor of a two-pole-three phase 28-V 0.63-hp 4500-r/min permanent-magnet ac machine is shown in Fig. 4.1-1. The magnets are samarium cobalt and the drive inverter is supplied from a 28-V dc source. The magnetic end cap is used in conjunction with Hall-effect sensors mounted in the stator housing (not shown) to determine the rotor position.

The majority of the electric power is generated by synchronous generators, that is, synchronous machines driven either by hydroturbines, steam turbines, or combustion engines. Just as the induction machine is the workhorse when it comes to converting energy from electric to mechanical, the three-phase synchronous generator is the principal means of converting energy from mechanical to electrical. As it turns out, we are able to analyze the synchronous machine using work we have done in earlier chapters. We will not repeat the derivation; instead, we will set forth the necessary equations using the equivalent circuits we have already developed.

A four-pole three-phase salient-pole synchronous machine is shown in Fig. 4.1-2. Note the dc machine connected to the shaft for purposes of supplying voltage to the

Introduction to Modern Analysis of Electric Machines and Drives, First Edition.
Paul C. Krause and Thomas C. Krause.
© 2023 The Institute of Electrical and Electronics Engineers, Inc.
Published 2023 by John Wiley & Sons, Inc.

Figure 4.1-1 Two-pole three-phase 28-V 0.63-hp 4500-r/min permanent-magnet ac machine. *Source:* Courtesy of Vickers Electromech.

Figure 4.1-2 Four-pole three-phase salient-pole synchronous machine.

field winding of the synchronous machine. Note also, the squirrel-cage damper windings embedded in the pole faces.

4.2 Analysis of the Permanent-Magnet ac Motor

The two-pole three-phase permanent-magnet ac or synchronous machine is shown in Fig. 4.2-1. The three-phase stator voltage equations are

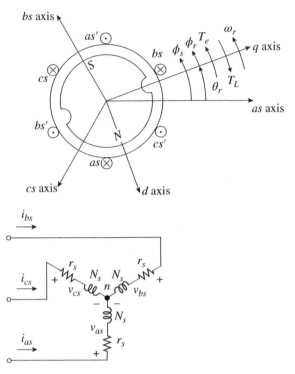

Figure 4.2-1 Two-pole three-phase permanent-magnet ac machine.

$$v_{as} = r_s i_{as} + \frac{d\lambda_{as}}{dt} \tag{4.2-1}$$

$$v_{bs} = r_s i_{bs} + \frac{d\lambda_{bs}}{dt} \tag{4.2-2}$$

$$v_{cs} = r_s i_{cs} + \frac{d\lambda_{cs}}{dt} \tag{4.2-3}$$

Since the rotor is not electrically symmetrical, it is convenient to use the rotor reference frame. From (2.4-15) and (2.4-16), the voltage equations with v_{0s} added for the three-phase stator in the rotor reference frame are

$$v_{qs}^r = r_s i_{qs}^r + \omega_r \lambda_{ds}^r + p\lambda_{qs}^r \tag{4.2-4}$$

$$v_{ds}^r = r_s i_{ds}^r - \omega_r \lambda_{qs}^r + p\lambda_{ds}^r \tag{4.2-5}$$

$$v_{0s} = r_s i_{0s} + p\lambda_{0s} \tag{4.2-6}$$

We need λ_{qd0s} which we can obtain from λ_{abcs} which is expressed as

$$\lambda_{as} = L_{asas}i_{as} + L_{asbs}i_{bs} + L_{ascs}i_{cs} + \lambda_{asm} \tag{4.2-7}$$

$$\lambda_{bs} = L_{bsas}i_{as} + L_{bsbs}i_{bs} + L_{bscs}i_{cs} + \lambda_{bsm} \tag{4.2-8}$$

$$\lambda_{cs} = L_{csas}i_{as} + L_{csbs}i_{bs} + L_{cscs}i_{cs} + \lambda_{csm} \tag{4.2-9}$$

In matrix form,

$$\boldsymbol{\lambda}_{abcs} = \mathbf{L}_s \mathbf{i}_{abcs} + \boldsymbol{\lambda}'_m \tag{4.2-10}$$

where $\boldsymbol{\lambda}'_m$ is the flux linkages due to the permanent magnet of the rotor referred to the stator windings. From (4.2-7) through (4.2-9), $\boldsymbol{\lambda}'_m$ can be written as

$$\boldsymbol{\lambda}'_m = \begin{bmatrix} \lambda_{asm} \\ \lambda_{bsm} \\ \lambda_{csm} \end{bmatrix} = \lambda'_m \begin{bmatrix} \sin\theta_r \\ \sin\left(\theta_r - \dfrac{2}{3}\pi\right) \\ \sin\left(\theta_r + \dfrac{2}{3}\pi\right) \end{bmatrix} \tag{4.2-11}$$

where

$$\omega_r = \frac{d\theta_r}{dt} \tag{4.2-12}$$

The amplitude of $p\lambda_{asm}$, $p\lambda_{bsm}$, and $p\lambda_{csm}$ would be proportional to the open-circuited phase voltages if the rotor is driven at some rotational speed.

The inductance matrix $\mathbf{L}_s$ for a wye-connected stator is given by (1.5-30) and repeated here as

$$\mathbf{L}_s = \begin{bmatrix} L_{ls} + L_{ms} & -\dfrac{1}{2}L_{ms} & -\dfrac{1}{2}L_{ms} \\ -\dfrac{1}{2}L_{ms} & L_{ls} + L_{ms} & -\dfrac{1}{2}L_{ms} \\ -\dfrac{1}{2}L_{ms} & -\dfrac{1}{2}L_{ms} & L_{ls} + L_{ms} \end{bmatrix} \tag{4.2-13}$$

The transformation to the rotor reference frame is

$$\mathbf{f}^r_{qd0s} = \mathbf{K}^r_s \mathbf{f}_{abcs} \tag{4.2-14}$$

where

$$\left(\mathbf{f}^r_{qd0s}\right)^T = \begin{bmatrix} f^r_{qs} & f^r_{ds} & f_{0s} \end{bmatrix} \tag{4.2-15}$$

$$\left(\mathbf{f}_{abcs}\right)^T = \begin{bmatrix} f_{as} f_{bs} f_{cs} \end{bmatrix} \tag{4.2-16}$$

From (2.3-18) with $\theta = \theta_r$,

$$\mathbf{K}_s^r = \frac{2}{3} \begin{bmatrix} \cos\theta_r & \cos\left(\theta_r - \frac{2}{3}\pi\right) & \cos\left(\theta_r + \frac{2}{3}\pi\right) \\ \sin\theta_r & \sin\left(\theta_r - \frac{2}{3}\pi\right) & \sin\left(\theta_r + \frac{2}{3}\pi\right) \\ \frac{1}{2} & \frac{1}{2} & \frac{1}{2} \end{bmatrix} \tag{4.2-17}$$

where θ_r is defined by (4.2-12). The flux linkage λ_{qd0s} may be expressed as

$$\lambda_{qd0s} = \mathbf{K}_s^r \lambda_{abcs} \tag{4.2-18}$$

substituting (4.2-10) from λ_{abcs} gives

$$\lambda_{qd0s} = \mathbf{K}_s^r \mathbf{L}_s + \mathbf{K}_s^r \lambda_m' \tag{4.2-19}$$

which may be written as

$$\lambda_{qd0s}^r = \begin{bmatrix} L_{ls} + \frac{3}{2}L_{ms} & 0 & 0 \\ 0 & L_{ls} + \frac{3}{2}L_{ms} & 0 \\ 0 & 0 & L_{ls} \end{bmatrix} \begin{bmatrix} i_{qs}^r \\ i_{ds}^r \\ i_{0s} \end{bmatrix} + \lambda_m''^r \begin{bmatrix} 0 \\ 1 \\ 0 \end{bmatrix} \tag{4.2-20}$$

where $\frac{3}{2}L_{ms}$ will be replaced with L_{Ms} and a superscript r has been added to λ_m' to indicate that it is in the rotor reference frame; however, $\lambda_m''^r = \lambda_m'$. The first term of (4.2-20) comes from the stator windings which is given by (2.4-31). The second (last) term is due to the permanent magnet of the rotor. In expanded form, the flux-linkage equations become

$$\lambda_{qs}^r = L_{ss} i_{qs}^r \tag{4.2-21}$$

$$\lambda_{ds}^r = L_{ss} i_{ds}^r + \lambda_m''^r \tag{4.2-22}$$

$$\lambda_{0s} = L_{ls} i_{0s} \tag{4.2-23}$$

where

$$L_{ss} = L_{ls} + L_{Ms} \tag{4.2-24}$$

Equations (4.2-4) through (4.2-7) and (4.2-21) through (4.2-24) suggest the equivalent circuit shown in Fig. 4.2-2.

4.2.1 Torque

The expression for the electromagnetic torque may be obtained from a power balance approach. To do this, we must multiply (4.2-4) by i_{qs}^r and (4.2-5) by i_{ds}^r and add the results to obtain the total power input at the stator terminals. This is then

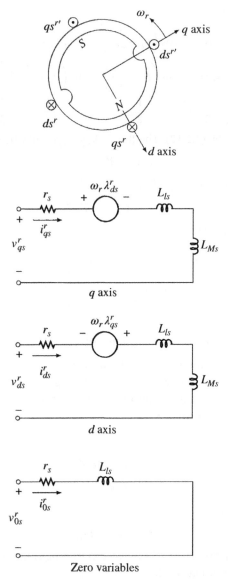

Figure 4.2-2 Equivalent circuit in rotor reference frame. $\lambda^r_{qs} = L_{ss}i^r_{qs}$, $\lambda^r_{ds} = L_{ss}i^r_{ds} + \lambda'^r_m$.

equated to the mechanically power which is $T_e \dfrac{2}{P} \omega_r$. Thus, accounting for Park's $\dfrac{2}{3}$ factor, we have

$$\frac{3}{2}\left(v_{qs}^r i_{qs}^r + v_{ds}^r i_{ds}^r\right) = T_e \frac{2}{P}\omega_r \tag{4.2-25}$$

which can be written as

$$\frac{3}{2}\left[r_s\left(i_{qs}^{r2} + i_{ds}^{r2}\right) + \omega_r\left(\lambda_{ds}^r i_{qs}^r - \lambda_{qs}^r i_{ds}^r\right) + p\left(\lambda_{qs}^r i_{qs}^r + \lambda_{ds}^r i_{ds}^r\right)\right]$$
$$= T_e \frac{2}{P}\omega_r \tag{4.2-26}$$

If we now equate coefficients of ω_r, we have

$$T_e = \frac{P}{2}\frac{3}{2}\left(\lambda_{ds}^r i_{qs}^r - \lambda_{qs}^r i_{ds}^r\right) \tag{4.2-27}$$

substituting in for λ_{qs}^r and λ_{ds}^r from (4.2-21) and (4.2-22), respectively, we have

$$T_e = \frac{P}{2}\frac{3}{2}\lambda_m^{\prime r} i_{qs}^r \tag{4.2-28}$$

Example 4.A Rotor Flux Linkage Referred to the Stator

The parameters of a four-pole three-phase permanent-magnet ac machine are $r_s = 3.4\,\Omega$, $L_{ls} = 1.1$ mH, and $L_{ms} = \dfrac{3}{2}11$ mH. When the device is driven at 1000 r/min, the open-circuit phase voltage is sinusoidal with a peak-to-peak value of 34.6 V. Determine λ_m'. The actual rotor speed at which the measurement was taken is

$$\omega_{rm} = \frac{(\text{r}/\min)(\text{rad}/\text{r})}{\text{s}/\min}$$
$$= \frac{(1000)(2\pi)}{60} = \frac{100}{3}\pi \text{ rad/s} \tag{4A-1}$$

The electrical angular velocity is

$$\omega_r = \frac{P}{2}\omega_{rm}$$
$$= \frac{4}{2}\frac{100\pi}{3} = \frac{200}{3}\pi \text{ rad/s} \tag{4A-2}$$

With the phases open-circuited, the stator currents are zero. Thus, from (4.2-1) and (4.2-7),

$$v_{as} = \frac{d\left(\lambda_m' \sin\theta_r\right)}{dt} = \lambda_m' \omega_r \cos\theta_r \tag{4A-3}$$

Now the peak-to-peak voltage is 34.6 V; hence, from (4A-3), with the peak-to-peak voltage divided by 2, we have

$$\frac{34.6}{2} = \lambda'_m \left(\frac{200}{3} \pi \right) \tag{4A-4}$$

Solving for λ'_m yields

$$\lambda'_m = \frac{(34.6)(3)}{(2)(200\pi)} = 0.0826 \text{ V·s/rad} \tag{4A-5}$$

4.2.2 Unequal Direct- and Quadrature-Axis Inductances

In our analysis of the permanent-magnet ac machine, we have assumed that the reluctance of the permanent magnet rotor is the same in the q and d axes. This assumption simplifies the analysis and allows us to portray the main operating modes of the brushless dc drive without significant error. There is, however, a difference in the q- and d-axis reluctances that produces a reluctance torque in addition to the main torque produced due to the interaction of the rotating magnetic field and the permanent-magnet rotor. Therefore, the influence of the magnetically unsymmetrical rotor on the voltage and torque expressions warrants some discussion. In order to prevent becoming involved with three-phase trigonometry, we will treat the two-phase machine and indicate the changes necessary for a three-phase machine.

A two-phase permanent-magnet ac machine with unequal q- and d-axis inductances is shown in Fig. 4.2-3.

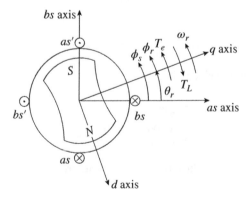

Figure 4.2-3 Two-phase permanent-magnet ac machine with unequal q- and d-axis reluctances, a salient-pole rotor.

The self-inductances of the stator windings and the mutual inductances between stator windings are functions of θ_r. Let us consider L_{asas}. With $\theta_r = 0$, we see from Fig. 4.2-3 that the magnetizing inductance of L_{asas} is less than it would be when $\theta_r = \frac{1}{2}\pi$. Let the magnetizing inductance of the as winding be denoted L_{mq} when $\theta_r = 0$, since in this position the q axis (high-reluctance path) is aligned with the magnetic axis of the as winding. Thus,

$$L_{asas} = L_{ls} + L_{mq}, \quad \text{when } \theta_r = 0 \tag{4.2-29}$$

where L_{ls} is the leakage inductance of the stator windings and

$$L_{mq} = \frac{N_s^2}{\mathcal{R}_{mq}} \tag{4.2-30}$$

where $\mathcal{R}_{mq}$ is an equivalent reluctance which is dominated by the two large air gaps of the magnetic path in the q axis. Now, at $\theta_r = \frac{1}{2}\pi$ the d axis (lower reluctance path which is dominated by the two smaller air gaps) is aligned with the magnetic axis of the as winding. Hence, denoting this magnetizing inductance as L_{md}, we can write

$$L_{asas} = L_{ls} + L_{md}, \quad \text{when } \theta_r = \frac{1}{2}\pi \tag{4.2-31}$$

where

$$L_{md} = \frac{N_s^2}{\mathcal{R}_{md}} \tag{4.2-32}$$

where $\mathcal{R}_{md}$ is an equivalent reluctance of the magnetic path in the d axis.

Since $\mathcal{R}_{mq} > \mathcal{R}_{md}$, $L_{mq} < L_{md}$, and we see that a minimum L_{asas} occurs at $\theta_r = 0$ and also again at $\theta_r = \pi$. In some permanent-magnet ac machines $\mathcal{R}_{mq} < \mathcal{R}_{md}$ and $L_{mq} > L_{md}$, due to the magnets in the d-axis; however, we will continue and take care of this later. Therefore, (4.2-29) is valid for $\theta_r = 0$ and π. Similarly, maximum L_{asas} occurs at $\theta_r = \frac{1}{2}\pi$ and again at $\theta_r = \frac{3}{2}\pi$; hence (4.2-31) applies for $\theta_r = \frac{1}{2}\pi$ and $\frac{3}{2}\pi$. The magnetizing inductance varies about an average value (which must be positive) and if we assume this variation to be sinusoidal, it would vary as a function of $2\theta_r$. Let L_A be the average value and L_B the amplitude of the sinusoidal variation about this average value. In this case,

$$L_{mq} = L_A - L_B \tag{4.2-33}$$

$$L_{md} = L_A + L_B \tag{4.2-34}$$

Substituting (4.2-30) and (4.2-32) for L_{mq} and L_{md}, respectively, into (4.2-33) and (4.2-34) and solving for L_A and L_B yields

$$L_A = \frac{N_s^2}{2}\left(\frac{1}{\mathcal{R}_{md}} + \frac{1}{\mathcal{R}_{mq}}\right) \tag{4.2-35}$$

$$L_B = \frac{N_s^2}{2}\left(\frac{1}{\mathcal{R}_{md}} - \frac{1}{\mathcal{R}_{mq}}\right) \tag{4.2-36}$$

Assuming a sinusoidal variation, we can write

$$L_{asas} = L_{ls} + L_A - L_B \cos 2\theta_r \tag{4.2-37}$$

If the air gaps were uniform as is the case in a round-rotor synchronous machine, $\mathcal{R}_{mq} = \mathcal{R}_{md}$ and, hence, from (4.2-36), $L_B = 0$.

By a similar procedure, it follows that, for the salient-pole device shown in Fig. 4.2-3,

$$L_{bsbs} = L_{ls} + L_A + L_B \cos 2\theta_r \tag{4.2-38}$$

Note when $\theta_r = 0$, L_{asas} is a minimum according to (4.2-37) and, according to (4.2-38), L_{bsbs} is a maximum. This, of course, corresponds to that which is portrayed in Fig. 4.2-3.

The mutual inductance $L_{asbs}(L_{bsas})$ is next. One would think that since the windings are orthogonal, the mutual coupling would always be zero; however, this is not the case due to the nonuniform air gap. Let us consider Fig. 4.2-4 where various rotor positions are shown with only the flux paths of the *as* winding depicted. Coupling occurs when flux produced by one winding links the other winding; for example, when the flux of the *as* winding links the *bs* winding. This will give us L_{bsas} and we know that $L_{asbs} = L_{bsas}$.

Note that, when $\theta_r = 0$, π, and 2π as shown in Fig. 4.2-4a or when $\theta_r = \frac{1}{2}\pi$ and $\frac{3}{2}\pi$ as shown in Fig. 4.2-4b. L_{bsas} (or L_{asbs}) is zero. In these positions, there is no channeling of the flux of one winding through the other. However, let the rotor start to turn counterclockwise from zero toward $\frac{1}{2}\pi$ and consider the flux produced by the *as* winding. As the rotor turns, the configuration of the rotor provides a low-reluctance path to the flux produced by the *as* winding and the flux is channeled through the *bs* winding with maximum coupling occurring at $\theta_r = \frac{1}{4}\pi$, as illustrated in Fig. 4.2-4c. We see that this same rotor position relative to the windings occurs also at $\theta_r = \frac{5}{4}\pi$. Maximum coupling will again occur at $\theta_r = \frac{3}{4}\pi$ and $\frac{7}{4}\pi$, as illustrated in Fig. 4.2-4c and d. Now, what is the sign of the mutual inductance? With the assumed direction of positive currents, the right-hand rule tells us that L_{bsas} (or L_{asbs}) is negative at $\theta_r = \frac{1}{4}\pi$, $\frac{5}{4}\pi$,... (the fluxes of the windings oppose each other for positive currents) and positive for

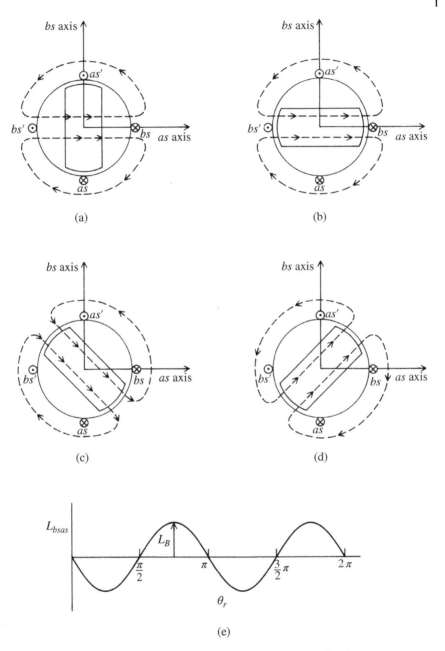

Figure 4.2-4 Flux path of *as* winding illustrating the mutual coupling between stator winding to determine L_{bsas} and L_{asbs}, (a) $\theta_r = 0, \pi$, and 2π; (b) $\theta_r = \frac{1}{2}\pi$ and $\frac{3}{2}\pi$; (c) $\theta_r = \frac{1}{4}\pi$ and $\frac{5}{4}\pi$; (d) $\theta_r = \frac{3}{4}\pi$ and $\frac{7}{4}\pi$; and (e) approximation of L_{bsas} and L_{asbs}.

$\theta_r = \dfrac{3}{4}\pi, \dfrac{7}{4}\pi,...$ (the fluxes aid each other). If we sketch L_{bsas} versus θ_r using the above information, we see, from Fig. 4.2-4e, that, as a first approximation, L_{bsas} or L_{asbs} may be expressed as

$$L_{bsas} = L_{asbs} = -L_B \sin 2\theta_r \tag{4.2-39}$$

In order for us to prove that the coefficient is L_B, it would be necessary to become quite involved [1]. We will accept this without proving it.

The expression for the stator self-inductance becomes

$$\mathbf{L_s} = \begin{bmatrix} L_{ls} + L_A - L_B \cos 2\theta_r & -L_B \sin 2\theta_r \\ -L_B \sin 2\theta_r & L_{ls} + L_A + L_B \cos 2\theta_r \end{bmatrix} \tag{4.2-40}$$

Now, in the rotor reference frame,

$$\begin{aligned} \begin{bmatrix} \lambda_{qs}^r \\ \lambda_{ds}^r \end{bmatrix} &= \mathbf{K_s^r L_s} \left(\mathbf{K_s^r}\right)^{-1} \mathbf{i}_{qds}^r + \mathbf{K_s^r} \lambda_m^{\prime r} \begin{bmatrix} \sin \theta_r \\ -\cos \theta_r \end{bmatrix} \\ &= \begin{bmatrix} L_q & 0 \\ 0 & L_d \end{bmatrix} \begin{bmatrix} i_{qs}^r \\ i_{ds}^r \end{bmatrix} + \lambda_m^{\prime r} \begin{bmatrix} 0 \\ 1 \end{bmatrix} \end{aligned} \tag{4.2-41}$$

In expanded form, (4.2-41) becomes

$$\lambda_{qs}^r = L_q i_{qs}^r \tag{4.2-42}$$

$$\lambda_{ds}^r = L_d i_{ds}^r + \lambda_m^{\prime r} \tag{4.2-43}$$

4.2.3 Three-Phase Machine

The equivalent circuits for the three-phase wye-connected machine with $L_q \neq L_d$ are the same as those given in Fig. 4.2-2 with L_{ms} replaced by L_{Mq} in the q axis and L_{Md} in the d axis where L_{Mq} is $\dfrac{3}{2}L_{mq}$ and L_{Md} is $\dfrac{3}{2}L_{md}$ and in (4.2-42) and (4.2-43), L_q is $L_{ls} + L_{Mq}$ and L_d is $L_{ls} + L_{Md}$.

The torque for the three-phase machine may be obtained from the equivalent circuits given in Fig. 4.2-2. The power $= T_e \omega_r +$ losses, therefore if we multiply v_{qs}^r by i_{qs}^r and v_{ds}^r by i_{ds}^r and equals coefficients of ω_r, we obtain T_e as

$$T_e = \left(\dfrac{3}{2}\right)\left(\dfrac{P}{2}\right)\left[\lambda_m^{\prime r} i_{qs}^r + \dfrac{3}{2}\left(L_d - L_q\right) i_{qs}^r i_{ds}^r\right] \tag{4.2-44}$$

where

$$L_q = L_{ls} + L_{Mq} \tag{4.2-45}$$

$$L_d = L_{ls} + L_{Md} \tag{4.2-46}$$

If your interest is in electric drives, you may skip to Chapters 5 and 6.

SP4.2-1. Express the mmf created by the permanent-magnet rotor $\left(\text{mmf}_r^r\right)$ as viewed from the rotor for the two-pole three-phase permanent-magnet ac machine shown in Fig. 4.2-1. Let F_p denote the peak value. [$\text{mmf}_r^r = -F_p \sin \phi_r$]

SP4.2-2. Obtain the last term of (4.2-20) from the last term of (4.2-19).

SP4.2-3. The inductive reactances for steady-state calculations are determined by multiplying the inductances by ω_r. Why? $[\omega_r = \omega_e]$

SP4.2-4. Determine the reluctance torque if $\tilde{I}_{as}$ is in phase with $\tilde{E}_a = |\tilde{E}_a| \underline{/0^\circ}$. [zero since $I_{ds}^r = 0$]

4.3 Windings of the Synchronous Machine

Before becoming involved in synchronous machine analysis, it is helpful to describe the function of the windings that are present in most synchronous machines. For this purpose, it is convenient to consider the windings of a two-pole three-phase salient-pole synchronous machine shown in Fig. 4.3-1. The stator windings are identical, sinusoidally distributed windings, as described in Chapter 2. For analysis purposes, the electrical characteristics of the rotor may be adequately represented with a field winding (*fd* winding) and short-circuited

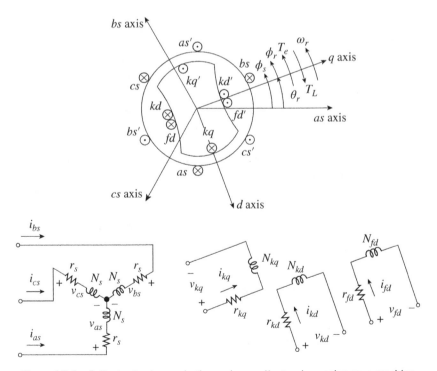

Figure 4.3-1 Salient-rotor two-pole three-phase salient-pole synchronous machine.

damper or *amortisseur* windings (*kq* and *kd* windings). Although the damper windings are shown with provisions to apply a voltage, they are, in fact, short-circuited windings which represent the paths for induced rotor currents. We will assume that the damper windings are approximated by two sinusoidally distributed windings displaced in space by 90° (see Chapter 3). The *kd* winding has the same magnetic axis as the *fd* winding, it has N_{kd} equivalent turns with resistance r_{kd}. The magnetic axis of *kq* winding is orthogonal with the magnetic axis of the *fd* and *kd* windings. It has N_{kq} equivalent turns and r_{kq} resistance. The rotor configuration shown in Fig. 4.3-1 for a three-phase machine is the same for any multi-phase synchronous machine. The quadrature axis (*q* axis) and direct axis (*d* axis) are also shown in Fig. 4.3-1. The *q* axis is the magnetic axis of the *kq* winding, whereas the *d* axis is the magnetic axis of the *fd* and *kd* windings. In synchronous machine analysis, but not in general, the *q* and *d* axes are reserved to denote the rotor magnetic axes since, over the years, they have been associated with the physical structure of the synchronous machine rotor quite independent of any transformation. However, as we have seen in the previous chapters, these need not be associated with any physical axes in general machine analysis.

With balanced steady-state stator currents, an air-gap mmf (mmf$_s$) is established that rotates about the air gap of a two-pole machine at ω_e, the angular velocity of the stator currents. We will assume that a dc voltage is applied to the *fd* winding by a brush and slip ring arrangement. The resulting field current i_{fd} establishes an air-gap mmf (mmf$_r$) which is fixed with respect to the rotor. We understand that the air-gap mmf (poles) established by the field winding must rotate at the same angular velocity as Tesla's rotating air-gap mmf established by the stator currents in order to produce a nonzero average electromagnetic torque during steady-state operation. Therefore, the rotor must rotate in synchronism with the air-gap mmf established by the stator windings ($\omega_r = \omega_e$); hence, the name synchronous machine. That is, the main torque production mechanism is this interaction of the air-gap mmf established by the stator currents mmf$_s$ and the air-gap mmf due to the direct current flowing in the field winding mmf$_r$. Albeit small, a reluctance torque is also developed at synchronous speed as a result of the saliency of the rotor. The so-called salient-pole construction is common for slower speed machines (large number of poles) such as hydroturbine generators. In this type of rotor construction, the field winding is wound upon the rotor surface, as shown in Fig. 4.3-1, and the air-gap is nonuniform to make room for the placement of the field winding. Therefore, the *q*-axis magnetic path has a higher reluctance than the *d*-axis magnetic path.

At this point, it seems that other than the addition of the damper windings and the saliency of the rotor, the important difference between the permanent-magnet ac machine considered in the previous section and this machine is that the permanent magnet is replaced with a field winding. Why then are we considering these two synchronous machines separately? The answer is in the mode of operation. The primary use of a permanent-magnet ac machine is in a brushless dc drive, where the frequency of the applied stator voltages is controlled to be the electrical angular velocity of the rotor. The voltages applied to the stator of the synchronous generator are fixed in frequency and magnitude by the power system. The synchronous generators in a power system are operating at the same electrical angular velocity (frequency). That is, they are all synchronized, which is necessary in order for each machine to supply power to the grid. We will talk more about this as we go along.

We have yet to discuss the function of the damper windings. It was found early on, that a synchronous machine with only a field winding would tend to oscillate about synchronous speed in a slowly damped manner following any slight disturbance. Adding damper windings (short-circuited rotor windings) provided the desired damping by induction machine action. The main torque of a synchronous machine is developed at synchronous speed because of the interaction of mmf_s and mmf_r. At synchronous speed, current is not induced in the damper windings and, hence, "induction machine" torque is not developed when $\omega_r = \omega_e$ (Chapter 3). If, however, for any reason the speed of the rotor should vary from synchronous speed, currents will be induced in the damper windings and the torque developed due to induction machine action, although small, it will damp oscillations of the rotor speed.

Torque is torque by whatever means it is developed and, perhaps, we should not emphasize the separation of torque into three types (interaction of mmf_s and mmf_r, reluctance, and induction) since the dynamic operation of the machine is described by nonlinear equations and superposition cannot be applied, in general. Nevertheless, this separation is helpful for our first look at the synchronous machine.

SP4.3-1. Express mmf_r for the two-pole three-phase synchronous machine shown in Fig. 4.3-1. $\left[\text{mmf}_r = -\left(N_{fd}/2\right)i_{fd}\sin\phi_r, \text{where } \phi_r \text{ is the counterclockwise displacement from the } fd \text{ axis}\right]$

SP4.3-2. The damper windings are short-circuited and the machine shown in Fig. 4.3-1 is driven at ω_r, counterclockwise. Assume that the stator currents are balanced 60-Hz currents with an acb-sequence. Determine the frequency of the currents flowing in the damper windings. $\left[\omega_r + \omega_e\right]$

4.4   Equivalent Circuit – Voltage and Torque Equations

Let us consider the synchronous machine shown in Fig 4.3-1. The stator is the same as that treated in Section 4.2 where L_q and L_d are not the same and a reluctance torque is produced. The damper windings, kq and kd, provide induction machine torque whenever the rotor electrical angular velocity, ω_r, is above or below synchronous speed. Although these windings are not symmetrical and the torque pulsates, the average value is similar to that shown in Fig. 3.8-2. The only thing that we have not considered in previous chapters is the field winding which actually is similar to the permanent-magnet ac machine considered in Section 4.2. In the case of the field winding, the rotor main flux is established by an adjustable field current rather than a permanent magnet.

Since we have already considered the stator and rotor circuits, we need not rederive the voltage equations. Instead, we can obtain the voltage equations from the equivalent circuits which we can construct from our previous work. This is shown in Fig. 4.4-1. The voltage equations in the rotor reference frame may be written as

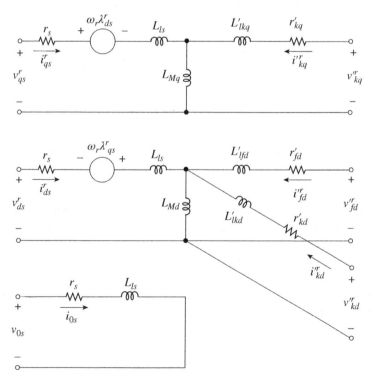

Figure 4.4-1   Equivalent circuit for three-phase synchronous machine in the rotor reference frame.

$$v_{qs}^r = r_s i_{qs}^r + \omega_r \lambda_{ds}^r + p\lambda_{qs}^r \tag{4.4-1}$$

$$v_{ds}^r = r_s i_{ds}^r - \omega_r \lambda_{qs}^r + p\lambda_{ds}^r \tag{4.4-2}$$

$$v_{0s} = r_s i_{0s} + p\lambda_{0s} \tag{4.4-3}$$

$$v_{kq}^{\prime r} = r_{kq}^\prime i_{kq}^{\prime r} + p\lambda_{kq}^{\prime r} \tag{4.4-4}$$

$$v_{fd}^{\prime r} = r_{fd}^\prime i_{fd}^{\prime r} + p\lambda_{fd}^{\prime r} \tag{4.4-5}$$

$$v_{kd}^{\prime r} = r_{kq}^\prime i_{kd}^{\prime r} + p\lambda_{kd}^{\prime r} \tag{4.4-6}$$

The flux-linkage equations for a three-phase stator become

$$\lambda_{qs}^r = L_{ls} i_{qs}^r + L_{Mq}\left(i_{qs}^r + i_{kq}^{\prime r} \right) \tag{4.4-7}$$

$$\lambda_{ds}^r = L_{ls} i_{ds}^r + L_{Md}\left(i_{ds}^r + i_{fd}^{\prime r} + i_{kd}^{\prime r} \right) \tag{4.4-8}$$

$$\lambda_{0s} = L_{ls} i_{0s} \tag{4.4-9}$$

$$\lambda_{kq}^{\prime r} = L_{lkq}^\prime i_{kq}^{\prime r} + L_{Mq}\left(i_{qs}^r + i_{kq}^{\prime r} \right) \tag{4.4-10}$$

$$\lambda_{fd}^{\prime r} = L_{lfd}^\prime i_{fd}^{\prime r} + L_{Md}\left(i_{ds}^r + i_{fd}^{\prime r} + i_{kd}^{\prime r} \right) \tag{4.4-11}$$

$$\lambda_{kd}^{\prime r} = L_{lkd}^\prime i_{kd}^{\prime r} + L_{Md}\left(i_{ds}^r + i_{fd}^{\prime r} + i_{kd}^{\prime r} \right) \tag{4.4-12}$$

where $L_{Mq} = \dfrac{3}{2}(L_A - L_B)$ and $L_{Md} = \dfrac{3}{2}(L_A + L_B)$.

In the voltage equations, the resistances, leakage inductances and currents of the rotor windings referred to the stator windings are

$$r_j^\prime = \frac{3}{2}\left(\frac{N_s}{N_j}\right)^2 r_j \tag{4.4-13}$$

$$L_{lj}^\prime = \frac{3}{2}\left(\frac{N_s}{N_j}\right)^2 L_{lj} \tag{4.4-14}$$

$$i_j^\prime = \left(\frac{2}{3}\right)\left(\frac{N_j}{N_s}\right) i_j \tag{4.4-15}$$

$$v_j^\prime = \left(\frac{N_s}{N_j}\right) v_j \tag{4.4-16}$$

$$\lambda_j^\prime = \left(\frac{N_s}{N_j}\right) \lambda_j \tag{4.4-17}$$

where j may be kq, fd, or kd.

4.4.1 Torque

We realize that the power output of a three-phase synchronous machine is

$$P_{out} = \left(\frac{2}{P}\right) T_e \omega_r \tag{4.4-18}$$

The power input is

$$P_{in} = \frac{3}{2}\left(v_{qs}^r i_{qs}^r + v_{ds}^r i_{ds}^r\right) \tag{4.4-19}$$

We see from the equivalent circuits given in Fig. 4.4-1, that the coefficients of ω_r are $\frac{3}{2}$ times $\lambda_{ds}^r i_{qs}^r$ and $-\lambda_{qs}^r i_{ds}^r$, thus setting those coefficients equal to $T_e \frac{2}{P}$ yields

$$T_e = \left(\frac{3}{2}\right)\left(\frac{P}{2}\right)\left(\lambda_{ds}^r i_{qs}^r - \lambda_{qs}^r i_{ds}^r\right) \tag{4.4-20}$$

Substituting (4.4-7) and (4.4-8) into (4.4-20) with $L_q = L_{ls} + L_{Mq}$ and $L_d = L_{ls} + L_{Md}$ yields

$$\begin{aligned}
T_e &= \left(\frac{3}{2}\right)\left(\frac{P}{2}\right)\left(L_d i_{ds}^r i_{qs}^r + L_{Md} i_{fd}''^r i_{qs}^r + L_{Md} i_{kq}''^r i_{qs}^r \right. \\
&\quad \left. - L_q i_{qs}^r i_{ds}^r + L_{Mq} i_{kq}''^r i_{ds}^r\right) \\
&= \left(\frac{3}{2}\right)\left(\frac{P}{2}\right)\left[L_{Md} i_{fd}''^r i_{qs}^r + (L_d - L_q) i_{qs}^r i_{ds}^r \right. \\
&\quad \left. + \left(L_{Md} i_{kd}''^r i_{qs}^r - L_{Mq} i_{kq}''^r i_{ds}^r\right)\right]
\end{aligned} \tag{4.4-21}$$

The first term inside the [ ] is the main torque produced by the field mmf$_r$ and the stator mmf$_s$. Note that in the case of the permanent ac machine $\lambda_m' = L_{Md} i_{fd}''^r$. The second term is the reluctance torque from (4.2-44), and the third and fourth terms are the induction machine torque. Now since L_{Md} and L_{Mq} are different and $i_{kd}''^r$ and $i_{kq}''^r$ do not form a balanced set, the torque will pulsate around an average value.

The torque–speed relation for motor operation is, with B_m neglected

$$T_e = J\left(\frac{2}{P}\right)\frac{d\omega_r}{dt} + T_L \tag{4.4-22}$$

4.4.2 Rotor Angle

In the case of the synchronous machine, the rotor angle is defined as

$$\delta = \theta_r - \theta_{esv}$$
$$= (\omega_r - \omega_e)t + \theta_r(0) - \theta_{esv}(0) \qquad (4.4-23)$$

For steady-state operation $\omega_r = \omega_e$ and the voltage is fixed at zero, $\theta_{esv}(0) = 0$ by the power system and

$$\delta = \theta_r(0) \qquad (4.4-24)$$

SP4.4-1. In some cases there are two damper windings in the q-axis. Write the voltage equation for this winding. $[v'_{kq2} = r'_{kq2}i'_{kq2} + p\lambda'_{kq2}]$

SP4.4-2. Why is $\theta_{esv}(0)$ fixed at zero in (4.4-23)? [The large power system fixed the voltage applied to the machine.]

4.5 Dynamic and Steady-State Performances

It is instructive to observe the variables of the synchronous machine during dynamic and steady-state operation. For this purpose, the operation of a two-phase synchronous generator is illustrated by computer traces. Two-phase machines are not often used in practice; instead, three-phase machines are more common. Nevertheless, our purpose is to understand the theory and principles of operation of a synchronous machine. A two-phase machine is just as applicable in this regard as a three-phase machine.

The two-phase synchronous machine which we will consider is a four-pole 150-hp 440-V (rms) 60-Hz machine with the following parameters: $r_s = 0.2\Omega$, $L_{ls} = 1.14$mH, $L_{mq} = 11$mH, $L_{md} = 13.7$mH, $r'_{fd} = 0.013\Omega$, and $L'_{lfd} = 2.1$mH. The inertia of the rotor and connected mechanical load is $J = 16.6$ kg·m^2 and B_m is neglected. The machine is a salient-pole device equipped with damper windings.

The dynamic performance of this synchronous machine during a step decrease in load torque from zero to -400 N·m is illustrated in Fig. 4.5-1. Since this is generator operation, perhaps it is more appropriate to consider this as a step increase in input torque from zero to 400 N·m. In any event, the machine is initially operating at synchronous speed with the field voltage adjusted so that the open-circuit voltage of the stator windings is equal to the rated voltage of the machine (440 V). Therefore, the stator currents are very small since $T_L = 0$. Plotted are v_{as}, i_{as}, v_{bs}, i_{bs}, T_e, ω_r (electrical angular velocity), δ, and T_L.

Immediately upon the application of the input torque ($-T_L$), the machine accelerates above synchronous speed as shown in Fig. 4.5-1 and the rotor angle, δ, increases since $\delta = \theta_r - \theta_{esv}$. The rotor continues to speed up until the accelerating torque on the rotor is zero. This occurs when T_e is equal in magnitude to the input torque. As noted in Fig. 4.5-1, this occurs when the speed has increased to

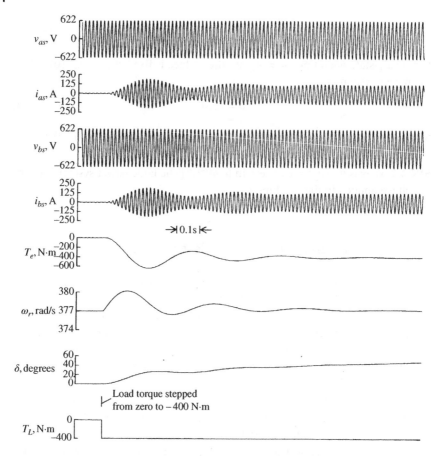

Figure 4.5-1 Dynamic performance of a two-phase synchronous generator during a step decrease in load torque (step increase in input torque).

approximately 380 rad/s (electrical angular velocity). Even though the accelerating torque on the rotor is zero at this time, the rotor is still running above synchronous speed. Hence, δ will continue to increase and, consequently, T_e will continue to decrease (increase negatively) causing the rotor to decelerate and the speed of the rotor decreases toward synchronous speed. When ω_r becomes equal to ω_e, the rotor angle is approximately 28 electrical degrees and T_e is approximately $-600 \, \text{N} \cdot \text{m}$. Now, there is a decelerating torque of approximately $200 \, \text{N} \cdot \text{m}$ causing the rotor speed to decrease below synchronous speed, whereupon the rotor angle will begin to decrease. Damped oscillations of the rotor about synchronous speed continue due to the damper windings until the new steady-state operating point is reached. We can think of the instantaneous electromagnetic torque during this disturbance resulting from interaction between (1) the stator and field currents, (2) the stator currents and saliency of the rotor, and (3) the stator and damper winding currents. Although this line of thinking may be helpful in visualizing what is going on, we must be careful since dynamically the device is an interactive, nonlinear, system.

The dynamic torque versus rotor angle characteristics during and following this step change in input torque are shown in Fig. 4.5-2. It is interesting to note that it requires considerable time before the machine establishes steady-state operation at $T_L = -400\,\text{N}\cdot\text{m}$. The steady-state torque–angle curve which is also shown, in part, in Fig. 4.5-2 will pass through $T_e = 0$ at $\delta \cong 0$ and $T_L = -400\,\text{N}\cdot\text{m}$ at $\delta = 68°$; however, it is important to note that the steady-state T_e versus δ characteristic is much different than the transient T_e versus δ characteristic. This is especially true on the first swing of the rotor where for $\delta = 20°$ the transient torque is over three times larger negatively than the steady-state torque.

If we slowly increase the input torque in small increments, theoretically we could reach the maximum steady-state value of T_e before the machine would fall out of synchronism. Since the dynamic characteristics are predicted by nonlinear differential equations, it is necessary to employ a computer to predict the dynamic

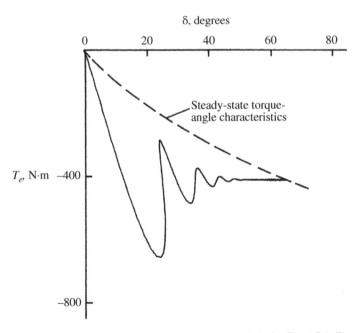

Figure 4.5-2 Dynamic torque versus rotor-angle characteristic for Fig. 4.5-1. The steady-state characteristics are also shown.

torque-angle characteristics. However, our purpose here is to make the first-time reader aware that the steady-state and dynamic torque versus rotor-angle characteristics are different, sometimes markedly different as illustrated here.

In Fig. 4.5-3, the rotor reference frame variables plotted rather than the stator or machine variables and the plot of the load torque omitted. Also plotted is the field current i''^r_{fd}. Although, for this machine, the field current changes only slightly owing to a change in flux linkages, this is not typical of all machines. In some cases, depending upon the parameters and the type of the disturbance, a considerable voltage may be induced in the field winding resulting in a change in field current during the transient period [1].

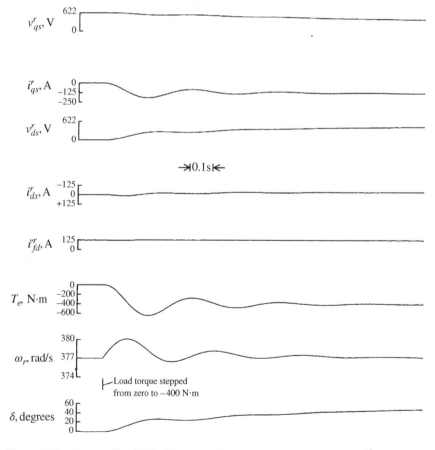

Figure 4.5-3 Same as Fig. 4.5-1 with rotor reference frame variables plotted, i''^r_{fd} added and T_L removed.

SP4.5-1. What changes would occur in T_e, i^r_{qs}, and δ in Fig. 4.5-3 if $T_L = 400\,\text{N} \cdot \text{m}$ rather than $-400\,\text{N} \cdot \text{m}$. [$T_e$ would become positive, i^r_{qs} would become positive, and δ would become negative.]

4.6 Analysis of Steady-State Operation

In the case of the synchronous machine, we are using the rotor reference frame. During steady-state operation, the electrical angular velocity of the rotor, ω_r, is equal to ω_e. Hence, the circuits that exist in the rotor reference frame are the field winding and the fictitious windings ($\acute{q}s$ and $\acute{d}s$ windings). These windings rotate at ω_r and therefore do not experience a change of flux linkages when $\omega_r = \omega_e$ as a result of relative motion between windings. Moreover, since i''_{fd} is constant and the rotor poles are shaped so that mmf^r_r is a constant amplitude sinusoidal function of ϕ_r and for balanced steady-state conditions, the stator windings are arranged so that mmf^e_s is also a constant amplitude sinusoidal function of ϕ_e. Therefore, there can be no induced voltage due to transformer action in any of the circuits in the rotor reference frame. One would then guess that the substitute currents and voltages associated with all windings in the rotor reference frame would be constant (zero in the case of the damper windings) during balanced steady-state operation. This seems logical, but what has happened to the balanced, sinusoidal stator variables? Remember the balanced, steady-state sinusoidal stator currents give rise to a constant amplitude mmf^e_s rotating at ω_e (i.e. Tesla's rotating magnetic field). But, if this constant amplitude mmf^e_s is now to be produced by substitute currents flowing in fictitious windings (i^r_{qs} and i^r_{ds}) that are mathematically fixed in the rotor reference frame, which rotates at ω_e during steady-state operation, these substitute currents must be constant.

It should be pointed out that during steady-state operation, we can relate F^r_{qs} and F^r_{ds} to $\tilde{F}_{as}$; however, when ω_r differs from ω_e, voltages are induced in the rotor windings that are of $\omega_e - \omega_r$ angular velocity. The synchronous reference frame variables will then contain multiple frequencies and the instantaneous phasor concept is not valid for sustained rotor speed different from ω_e. Although we will not set forth the derivation, we will use the instantaneous phasor later for small, temporary changes in rotor speed from synchronous.

Now that we know what to expect, let us proceed. During balanced steady-state operation, the stator variables may be expressed as

$$F_{as} = \sqrt{2}F_s \cos\left[\omega_e t + \theta_{esf}(0)\right] \tag{4.6-1}$$

$$F_{bs} = \sqrt{2}F_s \cos\left[\omega_e t - \frac{2\pi}{3} + \theta_{esf}(0)\right] \tag{4.6-2}$$

$$F_{cs} = \sqrt{2}F_s \cos\left[\omega_e t + \frac{2\pi}{3} + \theta_{esf}(0)\right] \tag{4.6-3}$$

Substituting (4.6-1) through (4.6-3) into the equations of transformation, (2.3-18), with $\theta = \theta_r$,

$$\theta_r = \omega_e t + \theta_r(0) \tag{4.6-4}$$

Thus,

$$F_{qs}^r = \sqrt{2}F_s \cos\left[\theta_{esf}(0) - \theta_r(0)\right] \tag{4.6-5}$$

$$F_{ds}^r = -\sqrt{2}F_s \sin\left[\theta_{esf}(0) - \theta_r(0)\right] \tag{4.6-6}$$

Since, $\theta_{esf}(0)$ is constant and $\theta_r(0)$ is $\delta(0)$, F_{qs}^r and F_{ds}^r are constants.

During steady-state balanced conditions, the stator and the field windings are the only windings carrying current. The voltage equations become

$$V_{qs}^r = r_s I_{qs}^r + \omega_r \lambda_{ds}^r \tag{4.6-7}$$

$$V_{ds}^r = r_s I_{ds}^r - \omega_r \lambda_{qs}^r \tag{4.6-8}$$

$$V_{fd}^{\prime r} = r_{fd}^{\prime} I_{fd}^{\prime r} \tag{4.6-9}$$

where

$$\lambda_{qs}^r = L_{ls} I_{qs}^r + L_{Mq} I_{qs}^r = L_q I_{qs}^r \tag{4.6-10}$$

$$\lambda_{ds}^r = L_{ls} I_{ds}^r + L_{Md}\left(I_{ds}^r + I_{fd}^{\prime r}\right) = L_d I_{ds}^r + L_{Md} I_{fd}^{\prime r} \tag{4.6-11}$$

where $L_q = L_{ls} + L_{Mq}$ and $L_d = L_{ls} + L_{Md}$.

Substituting $\theta_r(0)$ from (4.4-23) into (4.6-5) and (4.6-6) yields

$$F_{qs}^r = \sqrt{2}F_s \cos\left[\theta_{esf}(0) - \delta\right] \tag{4.6-12}$$

$$F_{ds}^r = -\sqrt{2}F_s \sin\left[\theta_{esf}(0) - \delta\right] \tag{4.6-13}$$

Now, $\tilde{F}_{as}$ is

$$\tilde{F}_{as} = F_s e^{j\theta_{esf}(0)} \tag{4.6-14}$$

If we multiply (4.6-14) by $\sqrt{2}e^{-j\delta}$, we can write

$$\sqrt{2}\tilde{F}_{as}e^{-j\delta} = \sqrt{2}F_s \cos\left[\theta_{esf}(0) - \delta\right] + j\sqrt{2}F_s \sin\left[\theta_{esf}(0) - \delta\right] \tag{4.6-15}$$

Substituting (4.6-12) and (4.6-13) into (4.6-15), we can write

$$\sqrt{2}\tilde{F}_{as}e^{-j\delta} = F_{qs}^r - jF_{ds}^r \tag{4.6-16}$$

Substituting (4.6-7) and (4.6-8) into (4.6-16) and since $\theta_{esv}(0) = 0$ we have

$$\sqrt{2}\tilde{V}_{as}e^{-j\delta} = r_s I_{qs}^r + X_d I_{ds}^r + \omega_e L_{Md} I_{fd}''^r + j\left(-r_s I_{ds}^r + X_q I_{qs}^r\right) \tag{4.6-17}$$

where $X = \omega_e L$ and

$$j\sqrt{2}\tilde{I}_{as}e^{-j\delta} = I_{ds}^r + jI_{qs}^r \tag{4.6-18}$$

In (4.6-17),

$$X_q = X_{ls} + X_{Mq} \tag{4.6-19}$$

$$X_d = X_{ls} + X_{Md} \tag{4.6-20}$$

If we add and substitute $X_q I_{ds}^r$ from the right-hand side of (4.6-17), we can write it as

$$\tilde{V}_{as} = \left(r_s + jX_q\right)\tilde{I}_{as} + \frac{1}{\sqrt{2}}\left[(X_d - X_q)I_{ds}^r + X_{Md}I_{fd}''^r\right]e^{j\delta} \tag{4.6-21}$$

The last term of (4.6-21) may be written as

$$\tilde{E}_a = \frac{1}{\sqrt{2}}\left[(X_d - X_q)I_{ds}^r + X_{Md}I_{fd}''^r\right]e^{j\delta} \tag{4.6-22}$$

We can now write (4.6-21) as

$$\tilde{V}_{as} = \left(r_s + jX_q\right)\tilde{I}_{as} + \tilde{E}_a \tag{4.6-23}$$

The phasor diagram is shown in Fig. 4.6-1 for typical generator action. The rotor poles are "pushing" the stator poles counterclockwise.

From (4.4-21), the steady-state torque for balanced conditions with r_s neglected may be written (damper winding currents are zero) as

$$T_e = -\left(\frac{3}{2}\right)\left(\frac{P}{2}\right)\frac{1}{\omega_e}\left[\frac{X_{Md}I_{fd}''^r\sqrt{2}V_s}{X_d}\sin\delta + \frac{1}{2}\left(\frac{1}{X_q} - \frac{1}{X_d}\right)\left(\sqrt{2}V_s\right)^2\sin 2\delta\right] \tag{4.6-24}$$

SP4.6-1. Why did we multiply (4.6-14) by $e^{-j\delta}$? [The q and d axes have been moved ccw by δ, so we must move them both to zero for (4.6-16) to apply.]

SP4.6-2. Why can r_s be neglected in the case of a synchronous generator but not in the case of a brushless drive? [constant frequency versus variable frequency operation.]

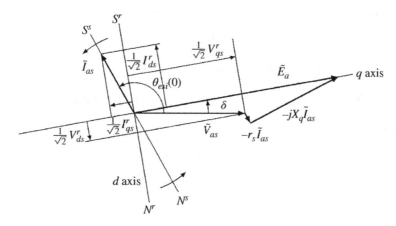

Figure 4.6-1 Phasor diagram for generator operation.

4.7 Transient Stability

Large-excursion stability, which is commonly referred to as transient stability, is of major concern to the power system engineer. Although we will not get too involved, we will describe some of the major aspects of this phenomenon. As we have mentioned, the vast majority of electric power is generated by synchronous generators and in order for multiple generators to supply power to an electric system, all of the generators must be synchronized with the electric system; in other words, all synchronous generators connected to an electric system must rotate at the same electrical angular velocity. Large areas of the power grid of the United States are "synchronously tied together." Since there are hundreds of synchronous generators connected to a large power grid, we can understand the concern of the power system engineer.

4.7.1 Three-Phase Fault

Our purpose is to describe large-excursion stability. To do this, we will consider the system in Fig. 4.7-1. Although this example may be considered academic, it serves our need without becoming overly involved. Initially, the steam turbine generator is delivering rated MVA (mega volt ampere) at rated power factor with essentially all of the power being delivered to the power grid.

The round-rotor machine parameters are

Rating: 835 MVA   Line-to-line voltage: 26 kV
Power factor: 0.85   Poles: 2   Speed: 3600 r/min
Inertia: $J = 0.0658 \times 10^6$ J · s^2 (turbine and generator)

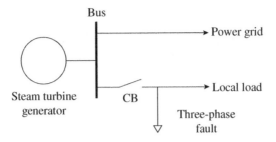

Figure 4.7-1 One-line diagram of system configuration for three-phase fault.

Resistances and reactances in ohms:

$$r_s = 2.43 \times 10^{-3} \qquad X_{Ms} = 1.3032 \qquad r'_{fd} = 7.5 \times 10^{-4}$$
$$X_{ls} = 0.1538 \qquad\qquad\qquad\qquad X'_{lfd} = 0.1145$$

A three-phase fault occurs on the line feeding the negligibly small local load near the generator. This is simulated by stepping the three-phase voltages at the bus to zero. The voltages are held at zero until the circuit breaker, CB, opens, clearing the fault from the bus. Whereupon it is assumed that the voltage at the bus steps back to normal magnitude, phase, and frequency due to the robustness or "stiffness" of the large system.

Figures 4.7-2 and 4.7-3 illustrate the dynamic performance of the steam turbine generator during and following the three-phase fault. There are a host of things that we need to explain regarding these computer traces. In Fig. 4.7-2, i_{as}, T_e, ω_r, and δ, the power angle, are plotted. The initial δ is $\delta(0) = 38.1°$.

The dynamic torque-angle characteristics, with the electric transients of the stator and transmission line neglected, during and following the fault, are illustrated in Fig. 4.7-3. The generator is initially operating in the steady state with the torque input $T_I = (0.85)2.22 \times 10^6 \text{N·m}$ and $E'_x = (2.48)\sqrt{\frac{2}{3}}26 \text{ kV}$. Recall we are now power system engineers and we changed the assumed direction of positive current; thus, the input torque is T_I and it is positive for generator action. With this change, stable steady-state operation occurs on the positive-slope part of the T_e versus δ plot $\left(-\frac{\pi}{2} < \delta < \frac{\pi}{2}\right)$; however, positive T_e occurs from $0 < \delta < \pi$.

In the more sophisticated present-day transient stability programs, machines are represented in their rotor reference frame and the network represented in the synchronously rotating reference frame. The electric transients are neglected in the stator windings of the machines and in the network; thereby, eliminating the need

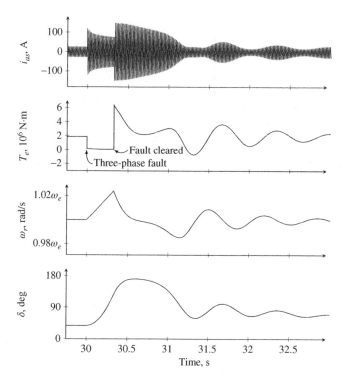

Figure 4.7-2 Dynamic performance of the steam turbine generator during a three-phase fault at the terminals predicted with stator and transmission line electric transients neglected.

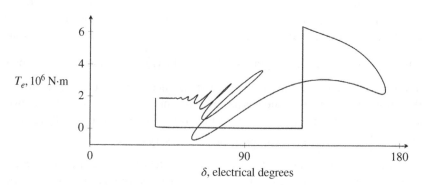

Figure 4.7-3 Torque versus angle characteristics for the study shown in Fig. 4.7-2.

to integrate for each inductance in the stator and network circuit equations while maintaining sufficient accuracy for transient stability studies. If the electric transients were taken into account, a rapidly decaying dc offset would be present in stator and network currents following an electrical disturbance. The decaying dc offset in the stator currents gives rise to a short-term 60-Hz decaying pulsation in T_e [1]. Although the damper windings are included in the simulation of the steam turbine generator for studies shown in Figs. 4.7-2 and 4.7-3, we will not include these windings in our following analysis since the effects of damper windings on transient stability is minimal.

Now, (4.4-23) defines δ and as we have mentioned, the power system is large and is often considered to have infinite electrical and mechanical inertias. Although the generator connected to the power system bus can deliver power to the system, the voltage and frequency of a large power system is essentially independent of one generator operation. This large power system bus is often referred to as an "infinite bus." Now, it can be argued that no such thing exists; clearly, a microgrid (small power system with few generators and loads) is not an infinite inertia system; nevertheless, we will assume an infinite bus in this discussion whereupon θ_{esv} is constant and since we are now power system engineers, (4.4-22) becomes

$$T_I - T_e = J\left(\frac{2}{P}\right)\frac{d^2\delta}{dt^2} \qquad (4.7\text{-}1)$$

since positive current is out of the machine and B_m is neglected. When the input torque, T_I, from the steam turbine is greater than the output torque, T_e, or power output, the machine accelerates; when $T_e > T_I$, it slows. Equation (4.7-1) is often referred to as the "swing equation." It is the dynamic relationship between the "swing" of the rotor of a generator relative to a large power system. That is, each generator connected to the power system would have its own swing equation made up of its own T_e, T_I, inertia, and rotor angle δ.

Let us now get back to Figs. 4.7-2 and 4.7-3. As mentioned, the three-phase fault is simulated by stepping the terminal voltages to zero; therefore, power cannot be transferred from the generator to the power grid. Thus, the input torque T_I accelerates the rotor as shown in the plot of ω_r in Fig. 4.7-2 where ω_e, in the plot of ω_r, is the electrical angular velocity of the power system. The fault is removed by opening the circuit breakers to the small local load in 0.334 seconds, whereupon the machine is reconnected to the power system and the oscillations in rotor speed subside due to the action of the damper windings. Had the fault been allowed to remain slightly longer and T_e had become less than T_I, the machine would again accelerate and would not have returned to synchronous operation with the grid without going through a time consuming resynchronization, or by what is called "slipping poles," which is not a common practice since system and/or machine damage can result.

Now, we should mention that a sudden three-phase fault is rare and a "bolted" three-phase fault near the machine is even rarer; however, it is often used in transient stability studies since it is a measure of the robustness of the system to faults. The dynamic torque-angle characteristics due to the three-phase fault shown in Fig. 4.7-3 are shown in the plot of T_e versus δ from Fig. 4.7-2. Now, we understand that the torque or power from the machine is essentially zero during the fault since the voltage is zero. The goal is to clear the fault and reconnect the machine to the grid and maintain synchronization of the machine with the grid. As we mentioned, synchronism is maintained if the fault is isolated by opening the circuit breaker feeding the fault in 0.334 seconds. Thus, the so-called "critical-clearing time" is 0.334 seconds and the "critical-clearing angle" which is the rotor displacement at the critical-clearing rotor angle is approximately 128 electrical degrees.

Let us "fast-backwards" to the early twentieth century when power systems were being built and the electric industry was growing rapidly and yet computers were years away. It appears that even at the start of World War II there were less than 20 integrators available in the United States and these were mechanical. Nevertheless, with only slide rules, the engineers were faced with the problem of predicting transient stability. Fortunately, R.E. Doherty and C.A. Nickle derived an approximate transient torque-angle characteristic that was sufficiently accurate when used with "equal-area criterion" to predict transient stability [1]. We are not going to dwell on a technique that has long since given way to modern-day computation; however, it is interesting that their method is based on the fact that during the "first swing" of the rotor, which determines stable or unstable operation, the flux linkages of circuits that are largely inductive with a relative small resistance, will tend to remain constant. Starting with this assumption they derived a transient torque–angle curve for the first rotor swing that was easy to determine and use. We should also mention in passing that during the 1930s and 1940s, point-to-point integration for transient stability studies was conducted using a "network calculator" which was a physical mimic of the main high-voltage part of the power system. This was an inconvenient time-consuming device that rapidly gave way to the digital computer in the 1960s.

Example 4.B Calculation of ω_r and δ at the Time of Fault Clearing

Consider Figs. 4.7-2 and 4.7-3. Calculate the rotor speed, ω_r, and rotor angle, δ, at the clearing time t_c of 0.334 seconds after the occurrence of the three-phase bolted fault when the synchronous generator is reconnected to the system. The input torque is

$$T_I = (0.85)2.22 \times 10^6 = 1.89 \times 10^6 \text{ N} \cdot \text{m} \tag{4B-1}$$

This input shaft torque remains on the generator during the duration of the fault, 0.334 seconds. For generator action, the power engineers write

$$T_I - T_e = J \left(\frac{2}{P}\right) \frac{d\omega_r}{dt} \tag{4B-2}$$

where T_L has become $-T_I$, T_e has become positive for generation action, and B_m is neglected. Solving for ω_r yields

$$\omega_r = \frac{P}{2J} \int (T_I - T_e) dt$$
$$= \frac{P}{2J} \int_0^{t_c} (T_I - T_e) d\xi \tag{4B-3}$$

where t_c is the clearing time, $J = 0.0658 \times 10^6$, and $P = 2$. Now, T_e is essentially zero during the time the fault is on the system; thus, (4B-3) becomes

$$\omega_r = \frac{(2)(1.89 \times 10^6)}{(2)(0.0658 \times 10^6)} \xi\big|_0^{0.344} + 377 = 9.59 + 377 = 386.6 \, \text{rad/s} \tag{4B-4}$$

This is $1.025 \, \omega_e$.

Now, (4.7-1) may be written as

$$\frac{d^2\delta}{dt^2} = \frac{2J}{P}(T_I - T_e) \tag{4B-5}$$

The initial rotor angle is $\delta(0) = 38.1°$ and during the fault $T_e = 0$; thus,

$$\delta = \frac{P}{2J} \int_0^{t_c} \int_0^t T_I \, dt \, d\tau + \delta(0)$$
$$= \left(\frac{P}{2J}\right)(T_I) \int_0^t \xi \, d\xi + \delta(0)$$
$$= \left[\left(\frac{P}{2J}\right) \frac{T_I}{2} \xi^2 \big|_0^{0.334}\right] \frac{360°}{2\pi} + \delta(0)$$
$$= \left[\frac{2(1.89 \times 10^6)(0.334)^2}{(2)(0.0658 \times 10^6)(2)}\right] \frac{360°}{2\pi} + 38.1° = 129.9° \tag{4B-6}$$

Note that t_c is the clearing time and that the $\dfrac{360°}{2\pi}$ multiplier is to convert radians to degrees.

SP4.7-1. Why is T_e during the three-phase not exactly zero. $\left[3 |\tilde{I}_{as}|^2 r_s\right]$

SP4.7-2. Assume $|I_{as}| = 40 \, \text{kA}$ during the fault. Calculate T_e. $[0.031 \times 10^6 \, \text{N} \cdot \text{m}]$

4.8 Problems

1 It is found that $\lambda''_m = 0.1$V·s/rad for a permanent-magnet six-pole two-phase ac machine. Calculate the amplitude (peak value) of the open-circuit phase voltage measured when the rotor is turned at 60 revolutions per second (r/s).

2 Consider the steam turbine generator in Section 4.7. Assume it is operating as a motor with rated power input and rated leading power factor. Draw the phasor diagram with all the quantities that are shown in Fig. 4.6-1.

3 Repeat Problem 2 for motor action with rated power factor lagging.

4 A four-pole 2-hp two-phase round-rotor synchronous machine is connected to a 110-V, 60-Hz source. The machine is operating as a generator with a total steady-state power output of 1 kW at the terminals. The phase current lags the phase voltage by 160°. The parameters are $r_s = 0.5\,\Omega$, $L_{ls} = 0.005$ H, and $L_{ms} = 0.05$ H. Calculate $\tilde{E}_a$ and draw the phasor diagram.

5 Repeat Problem 4 if the machine is operating as a motor with $\tilde{I}_{as}$ in phase with $\tilde{V}_{as}$. The input power is 1 kW.

6 Calculate the torque for Problem 5.

7 The field of a 60-Hz three-phase round-rotor synchronous machine is adjusted so that the open-circuited phase voltage is 14 kV. The parameters are $r_s = 0.003\,\Omega$, $X_{ls} = 0.1\,\Omega$, and $X_{ms} = 1.1\,\Omega$. Calculate i''^r_{fd}.

8 A three-phase synchronous generator is a 64-pole hydroturbine generator rated at 325 MVA with 20 kV line-to-line voltage. The machine delivers real power with a power factor of −0.85. The machine parameters in ohms at 60-Hz are $r_s = 0.003\,\Omega$, $X_{ls} = 0.1\,\Omega$, and $X_{Ms} = 1.2\,\Omega$. Calculate $\tilde{E}_a$, T_e, and draw the phasor diagram and locate the poles.

9 Obtain (4.6-24) from (4.4-21).

Reference

1 P. C. Krause, *Analysis of Electric Machinery*. McGraw-Hill Book Company, New York, 1986.

5

Direct Current Machine and Drive

5.1 Introduction

The direct-current (dc) machine is not as widely used today as it once was. The dc generator has been replaced by power electronics which convert alternating current into dc with provisions to control the magnitude of the dc voltage. In drive applications, the dc motor is being replaced by the voltage-controlled permanent-magnet ac machine (brushless dc drive) and/or the field-orientated induction motor. Although the analysis of a dc machine does not require a change of variables, it is still desirable to devote some time to the dc machine and dc drive since it is used as a low-power drive motor. There is another and perhaps more important reason to consider the dc machine. Although maintenance and environmental issues hamper the use of dc machines, this device is the only electric machine that is designed with the stator and rotor mmfs orthogonal, which inherently produces maximum torque per ampere. With the advent of power electronics, a huge effort is required to control the permanent-magnet ac and induction machines so as to emulate the characteristics of the dc motor. In this chapter, we will treat the dc machines sufficient to introduce the reader to the operating principles of dc machines with focus on the shunt-connected and permanent-magnet dc machine and drive. Thus, setting the stage for a comparison of the operating characteristics with the voltage-controlled permanent-magnet ac drive and the field-oriented induction motor drive set forth in Chapter 6.

A disassembled two-pole 0.1-hp 6-V 12,000-r/min permanent-magnet dc motor is shown in Fig. 5.1-1. The magnets, which replace the stator field winding, are samarium cobalt and the device is used to drive hand-held battery-operated surgical instruments. Although some of these terms are new to us, they will be defined as we go along.

Introduction to Modern Analysis of Electric Machines and Drives, First Edition.
Paul C. Krause and Thomas C. Krause.
© 2023 The Institute of Electrical and Electronics Engineers, Inc.
Published 2023 by John Wiley & Sons, Inc.

Figure 5.1-1 Two-pole 0.1-hp 6-V 12,000-r/min permanent-magnet dc motor. *Source:* Courtesy of Vick ElectroMech.

5.2 Commutation

An elementary dc machine is shown in cross section in Figs. 5.2-1 and 5.2-2. The field winding is carrying a dc i_f into the paper at f_1 and out at f_1' and then in at f_2 and out at f_2'. A voltage v_f is applied across f_1 and f_2'. With positive i_f, the field winding creates a mmf that is stationary and positive in the f axis.

The armature or rotor consists of two parallel windings, the a winding and the A winding. Each winding has four coils with each coil connected to two segments of the commutator. The commutator is fixed to the rotor and makes contact with carbon brushes. As the rotor rotates, the commutator segments slide against the brushes. This action connects the rotating circuits (a and A windings) to stationary terminals denoted as v_a which are connected to a dc source or to a load if the device is operating as a generator. Note that in Fig. 5.2-1 the dc current i_a is flowing into the top brush which is straddling two of the eight segments of the commutator. Each segment is insulated from the others. The top brush is short-circuiting the A_4 coil; the bottom brush is short-circuiting the a_4 coil. In Fig. 5.2-2, the brushes are not commutating any windings. Sinusoidal voltages are induced in each of the coils due to the constant field current or permanent magnet producing a stationary mmf_s and the windings rotating in this constant mmf_s. Due to the action of the commutator, the mmf_r (a axis) of the rotor is also essentially stationary and orthogonal to the field mmf_s (f axis).

The full-wave rectified voltages in Figs. 5.2-1 and 5.2-2 are the open-circuit voltage of one parallel path between brushes. This is referred to as the back voltage or back emf. This induced voltage exists only when the rotor is turning. In Figs. 5.2-1 and 5.2-2, the parallel windings each consist of four coils and produces an mmf that is orthogonal to the mmf produced by the stator winding, the f winding.

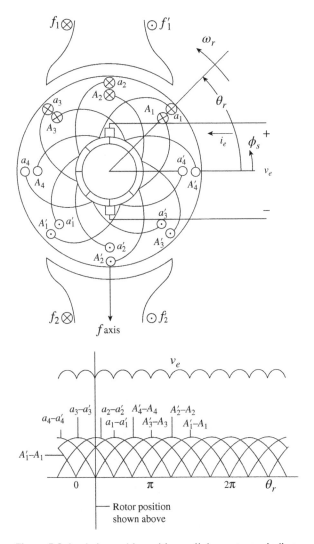

Figure 5.2-1 A dc machine with parallel armature windings.

It is important to understand that the commutator and brush combination is to change the direction of current flow in the rotor windings so that for positive current into the machine flows into the paper over the top part of the rotor and out over the bottom part for motor action. The current is reversed for generator action.

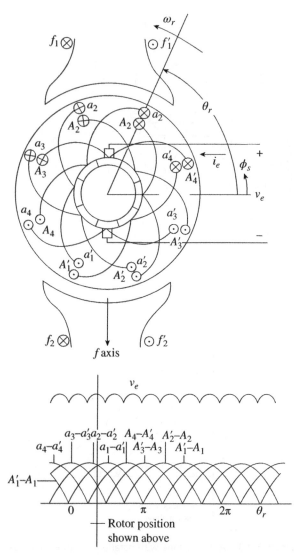

Figure 5.2-2 Same as Fig. 5.2-1 with rotor advanced approximately 22.5° counterclockwise.

SP5.2-1. The peak value of the voltage induced in one coil shown in Fig. 5.2-1 is 1 V. Determine, from Fig. 5.2-1, the maximum and minimum value of v_a. [2.813 V; 2.414 V]

5.3 Voltage and Torque Equations

It is advantageous to first consider the dc machine with a field and armature winding before turning to the permanent-magnet device exclusively. Although rigorous derivation of the voltage and torque equations is possible, it is rather lengthy and little is gained since these relationships may be deduced. The armature coils revolve in a stationary magnetic field established by a current flowing in the field winding. We have established that a voltage is induced in these coils by virtue of this rotation. However, the action of the commutator causes the armature coils to appear as a stationary winding with its magnetic axis orthogonal to the magnetic axis of the field winding. In other words, the stator and rotor mmf$_s$ are orthogonal. Therefore, voltages are not induced in one winding due to the time rate of change of the current flowing in the other (transformer action). Mindful of these conditions, we can write the field and armature voltage equations in matrix form as

$$\begin{bmatrix} v_f \\ v_a \end{bmatrix} = \begin{bmatrix} r_f + pL_{FF} & 0 \\ \omega_r L_{AF} & r_a + pL_{AA} \end{bmatrix} \begin{bmatrix} i_f \\ i_a \end{bmatrix} \tag{5.3-1}$$

where L_{FF} and L_{AA} are the self-inductances of the field and armature windings, respectively, and p is the short-hand notation for the operator d/dt. The rotor speed is denoted as ω_r, and L_{AF} is the mutual inductance between the field and the rotating armature coils which is readily determined from the open-circuited voltage. The above equation suggests the equivalent circuit shown in Fig. 5.3-1. The voltage induced in the armature circuit, $\omega_r L_{AF} i_f$, is commonly referred to as the counter or back emf. It also represents the open-circuit armature voltage from which L_{AF} can be readily determined. The equivalent circuit shown in Fig. 5.3-1 is for a separately excited machine where v_f is from a separate dc source. When the field and armature are connected to the same dc source, $v_f = v_a$, it is a shunt machine.

A substitute variable often used is

$$k_v = L_{AF} i_f \tag{5.3-2}$$

We will find that this substitute variable is particularly convenient and frequently used. Even though a permanent-magnet dc machine has no field circuit, the

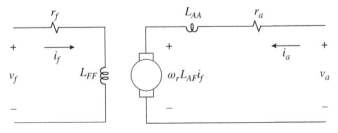

Figure 5.3-1 Equivalent circuit of dc machine.

constant field flux produced by the permanent magnet is analogous to a dc machine with a constant k_v.

The torque can be determined by first expressing mmf_s and mmf_a. From Figs. 5.2-1 and 5.2-2, one half of mmf_s and mmf_r becomes

$$\mathrm{mmf}_s = \frac{N_f}{2} \sin \phi_s i_f \tag{5.3-3}$$

$$\mathrm{mmf}_a = \frac{N_a}{2} \cos \phi_s i_a \tag{5.3-4}$$

The mmf_s are orthogonal. The field flux Φ_f may be expressed as $\dfrac{\mathrm{mmf}_s}{\mathcal{R}}$ where mmf_s is two times (5.3-3) and $\mathcal{R}$ is an equivalent reluctance. The torque may be expressed by the cross product of the maximum mmf_a, two times (5.3-4), and Φ_f. Thus,

$$
\begin{aligned}
T_e &= \left(\frac{N_f i_f}{\mathcal{R}}\right)(N_a i_a) \\
&= \frac{N_f N_a}{\mathcal{R}} i_f i_a \\
&= L_{AF} i_f i_a
\end{aligned}
\tag{5.3-5}
$$

The torque and rotor speed are related by

$$T_e = J \frac{d\omega_r}{dt} + B_m \omega_r + T_L \tag{5.3-6}$$

where J is the inertia of the rotor and rigidly connected mechanical load. The units of the inertia are $\mathrm{kg \cdot m^2}$ or $\mathrm{J \cdot s^2}$. A positive electromagnetic torque T_e acts to turn the rotor in the direction of increasing θ_r. The load torque T_L is positive for a torque, on the shaft of the rotor, which opposes the positive electromagnetic torque T_e. The constant B_m is a damping coefficient associated with the mechanical rotational system of the machine. It has the units of $\mathrm{N \cdot m \cdot s/rad}$ and it is generally small and often neglected.

Although we will focus on the permanent-magnet dc motor, it is worthwhile to take a moment to mention that we have established the basis for several types of dc machines. In particular, the machine shown in Fig. 5.3-1 is a separately excited dc machine. If we connect the field winding in parallel with the armature winding, it becomes a shunt-connected dc machine. If the field winding is connected in series with the armature winding, it is a series-connected dc machine. If two windings are used, one in parallel with and another in series with the armature, it is referred to as a compound-connected dc machine. Clearly, this is an overly simplistic description and the reader is referred to [1] for a more detailed consideration of these machine types.

Before proceeding, it is appropriate to mention briefly generator action even though the dc generator has been replaced by the ac to dc converter. We see from Fig. 5.3-1 that if $\omega_r L_{AF} i_f$ is greater than v_a, i_a is reversed and we have generator action. In this case, T_L is negative since the dc machine is being driven and T_e, (5.3-5) is negative and in the steady state, from (5.3-6) $\dfrac{d\omega_r}{dt}$ is zero and B_m is small.

SP5.3-1. The armature applied voltage is 240 V; the rotor speed is constant at 50 rad/s and $I_a = 15$ A. The armature resistance is 1 Ω and $L_{AF} = 1$ H. Calculate the steady-state field current. $\left[I_f = 4.5\text{ A}\right]$

5.4 Permanent-Magnet dc Machine

In the case of the permanent-magnet dc machine, $L_{AF}I_f$ is replaced with k_v, whereupon the steady-state armature voltage equation becomes

$$V_a = r_a I_a + k_v \omega_r \tag{5.4-1}$$

If (5.4-1) is solved for I_a and substituted into (5.3-5) with $L_{AF}I_f$ replaced by k_v, the steady-state torque may be expressed as

$$T_e = k_v I_a$$
$$= \frac{k_v V_a - k_v^2 \omega_r}{r_a} \tag{5.4-2}$$

The steady-state torque-speed characteristic is shown in Fig. 5.4-1.

It is apparent from Fig.5.4-1, that the stall ($\omega_r = 0$) torque could be made larger for a given armature voltage by reducing r_a. Although the machine may be designed with a smaller armature resistance, there is a problem since, at stall, the steady-state armature current is limited by the armature resistance; hence, for a constant V_a, reducing r_a will result in a larger I_a at stall which can cause

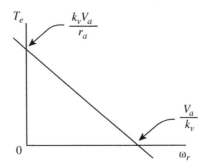

Figure 5.4-1 Steady-state torque-speed characteristic of a permanent-magnet dc machine.

damage to the brushes. On the other hand, increasing the starting torque by reducing r_a causes the torque-speed characteristics to have a steeper slope which results in a smaller change in speed for a given change in load torque during normal (near rated) operation. If, however, the armature voltage is reduced during the starting period to protect the brushes, the desirable characteristic of a small speed change for load torque variations during normal operation could be achieved. In fact, controlled regulation of the armature voltage is generally employed for large horsepower machines by using a converter; however, low-power permanent-magnet dc machines are generally supplied from a battery as in the case of the automobile and, therefore, a large armature resistance is necessary in order to prevent brush damage during the early part of the starting period. Fortunately, a small speed variation during load torque changes is not required in many applications of the permanent-magnet dc machine; therefore, the steep torque-speed characteristics are not necessary.

Example 5.A   Calculating Machine Parameters

A permanent-magnet dc motor is rated at 6 V with the following parameters: $r_a = 7\,\Omega$, $L_{AA} = 120\,\text{mH}$, $k_T = 2\,\text{oz} \cdot \text{in/A}$, and $J = 150\,\mu\,\text{oz} \cdot \text{in} \cdot \text{s}^2$. (a) Determine the stall torque and the no-load speed. (b) A torque load of 0.5 oz · in is applied, determine the steady-state ω_r. (c) Determine the efficiency at this load.

First, let us convert k_T and J to units which we have been using. In this regard, we will convert the inertia to $\text{kg} \cdot \text{m}^2$ which is the same as $\text{N} \cdot \text{m} \cdot \text{s}^2$. To do this, we must convert ounces to Newtons and inches to meters. Thus,

$$J = \frac{150 \times 10^{-6}}{(3.6)(39.37)} = 1.06 \times 10^{-6}\,\text{kg} \cdot \text{m}^2 \tag{5A-1}$$

We have not seen k_T before. It is the torque constant and, if expressed in the appropriate units, it is numerically equal to k_v. When k_v is used in the expression for T_e ($T_e = k_v i_a$), it is often referred to as the *torque constant* and denoted k_T. When used in the voltage equation, it is always denoted k_v. Now, we must convert oz into N m, whereupon k_T equals our k_v; hence,

$$k_v = \frac{2}{(3.6)(39.37)} = 1.41 \times 10^{-2}\,\text{N} \cdot \text{m/A} = 1.41 \times 10^{-2}\,\text{V} \cdot \text{s/rad} \tag{5A-2}$$

(a) The stall torque is

$$T_e = \frac{k_v V_a}{r_a} = \frac{(1.41 \times 10^{-2})(6)}{7} = 1.21 \times 10^{-2}\,\text{N} \cdot \text{m} \tag{5A-3}$$

The no-load speed is

$$\omega_r = \frac{V_a}{k_v} = \frac{6}{1.41 \times 10^{-2}} = 425.5 \text{ rad/s} \tag{5A-4}$$

$$T_L = \frac{0.5}{(3.6)(39.37)} = 3.5 \times 10^{-3} \text{ N} \cdot \text{m} \tag{5A-5}$$

(b) From (5.4-1) with $T_e = T_L$,

$$\omega_r = \frac{V_a}{k_v} - \frac{T_L r_a}{k_v^2} = \frac{6}{1.41 \times 10^{-2}} - \frac{(3.5 \times 10^{-3})(7)}{(1.41 \times 10^{-2})^2} \tag{5A-6}$$
$$= 425.5 - 123.2 = 302.3 \text{ rad/s}$$

(c) $T_e = k_v I_a$

$$3.5 \times 10^{-3} = (1.41 \times 10^{-2}) I_a$$

$$I_a = \frac{3.5 \times 10^{-3}}{1.41 \times 10^{-2}} = 0.248 \text{ A} \tag{5A-7}$$

$$P_{loss} = r_a I_a^2 = (7)(0.248)^2 = 0.431 \text{ W} \tag{5A-8}$$

$$P_{in} = V_a I_a = (6)(0.248) = 1.488 \text{ W} \tag{5A-9}$$

$$P_{out} = P_{in} - P_{loss} = 1.488 - 0.431 = 1.057 \text{ W} \tag{5A-10}$$

$$\text{Eff.} = \frac{P_{out}}{P_{in}} \times 100 = \frac{1.058}{1.488} \times 100 = 71\% \tag{5A-11}$$

Note that

$$P_{out} = T_e \omega_r = 3.5 \times 10^{-3} \times 302.3 = 1.058 \text{ W} \tag{5A-12}$$

which is essentially equal to (5A-10).

SP5.4-1. When a 12 V permanent-magnet dc motor is driven at 100 rad/s, the open-circuit voltage is 10 V. Calculate k_v. [$k_v = 0.1$ V·s/rad]

SP5.4-2. Multiply the expression for v_a given in (5.3-1) by i_a and identify all terms. [$v_a i_a$ – input power to armature, $i_a^2 r_a$ – armature ohmic loss, $L_{AA} i_a p i_a$ – change of energy stored in L_{AA}, $L_{AF} i_f i_a \omega_r = T_e \omega_r$ – output power]

SP5.4-3. Is SP5.4-2 an alternate method of deriving an expression for torque? [Yes; $L_{AF} i_f i_a$ is the coefficient of $p\theta_r$ or ω_r]

SP5.4-4. A 12-V permanent-magnet dc motor has an armature resistance of 12 Ω and $k_v = 0.01$ V·s/rad. Calculate the steady-state stall torque (T_e with $\omega_r = 0$). [$T_e = 0.01$ N·m].

5.5 DC Drive

Since the dc machine plays a role in some drive applications, a brief look at a voltage control drive is appropriate. Our focus will be on the permanent-magnet dc machine supplied from a two-quadrant dc converter. Dynamic and steady-state performances are illustrated. Since the dc converters used in dc drive systems are often called choppers, we will use dc converter and chopper interchangeably. In this section, we will analyze the operation and establish the average-value model for a two-quadrant chopper drive.

A two-quadrant dc converter is depicted in Fig. 5.5-1. The switches $S1$ and $S2$ are transistors. They are assumed to be ideal; that is, if $S1$ or $S2$ is closed, current is allowed to flow in the direction of the arrow; current is not permitted to flow opposite to the arrow. If $S1$ or $S2$ is open, current is not allowed to flow in either direction regardless of the voltage across the switch. If $S1$ or $S2$ is closed and the current is positive, the voltage drop across the switch is assumed to be zero. Similarly, the diodes $D1$ and $D2$ are ideal. Therefore, if the diode current i_{D1} or i_{D2} is greater than zero, the voltage across the diode is zero. The diode current can never be less than zero.

Waveforms of the converter variables during steady-state operation are shown in Fig. 5.5-2. Therein, the switching period T is made large relative to the armature time constant τ_a for the purpose of depicting the transient of the armature current. Normally, the switching period is much smaller than the armature time constant and the switching segments of i_a are essentially sawtooth in shape. This is portrayed later in this section. With a two-quadrant chopper, the armature voltage cannot be negative ($v_a \geq 0$); however, the armature current can be positive or negative. That is, I_1 and I_2 (Fig. 5.5-2) can both be positive, or I_1 can be negative and I_2 positive, or I_1 and I_2 can both be negative. In Fig. 5.5-2, I_1 is negative and I_2 is positive and the average value of i_a is positive. The mode of operation depicted is motor action if ω_r is positive (ccw).

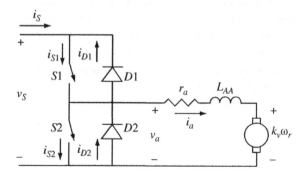

Figure 5.5-1 Two-quadrant chopper drive.

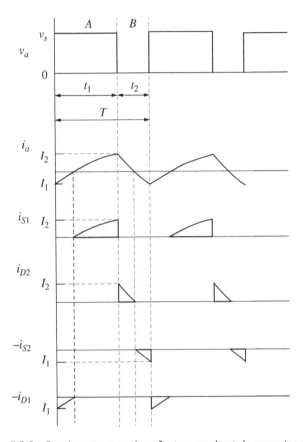

Figure 5.5-2 Steady-state operation of a two-quadrant dc converter drive.

During interval A, $S1$ is closed and $S2$ is open and, at the start of interval A, $i_a = I_1$, which is negative. Since $S2$ is open, a negative i_a (I_1) can only flow through $D1$. It is important to note that $-i_{D1}$ and $-i_{S2}$ are plotted in Fig. 5.5-2 to allow-ready comparison with the waveform of i_a, since they are opposite to positive i_a. Let us go back to the start of interval A. How did i_a become negative? Well, during the interval B in the preceding period, $S2$ was closed with $S1$ open. With $S2$ closed, the armature terminals are short-circuited and the counter emf has driven i_a neg-ative. Therefore, when $S1$ is closed and $S2$ is opened at the start of interval A, the source voltage has to contend with this negative I_1. We see from Fig. 5.5-2 that the average value of i_a is slightly positive; therefore, v_S is larger than the counter emf and at the start of interval A when v_S is applied to the machine, the armature cur-rent begins to increase toward zero from the negative value of I_1. Once i_a reaches zero, the diode D_1 blocks the current flow. That is, i_{D1} cannot become negative (cannot conduct positive i_a); however, $S1$ has been closed since the start of interval A and since i_{S1} can only be positive, $S1$ is ready to carry the positive i_a. The armature current, which is now i_{S1}, continues to increase until the end of interval A (I_2).

At the beginning of interval B, $S1$ is opened and $S2$ is closed; however, $S2$ cannot conduct a positive armature current. Therefore, the positive current (I_2) is diverted to diode $D2$ which is short-circuiting the armature terminals. Now, the counter emf has the positive current (I_2) with which to contend. It is clear that if the armature terminals were permanently short-circuited, the counter-emf would drive i_a negative. At the start of interval B, the counter-emf begins to do just that; however, when i_a becomes zero, diode $D2$ blocks i_{D2} and the negative armature current is picked up by $S2$, which has been closed since the beginning of interval B, waiting to be called upon to conduct a negative armature current. This continues until the end of interval B, whereupon we are back to where we started.

It is apparent that if the mode of operation is such that I_1 and I_2 are both positive, then the machine is acting as a motor with a substantial load torque if ω_r is positive (ccw). In this mode, either $S1$ or $D2$ will carry current during a switching period T. If both I_1 and I_2 are negative, the machine is operating as a generator, delivering power to the source if ω_r is driven ccw. In this case, either $S2$ or $D1$ will carry current during a switching period.

5.5.1 Average-Value Time-Domain Block Diagram

The average-value time-domain block diagram for the two-quadrant chopper drive system is shown in Fig. 5.5-3. From Fig. 5.5-2, the average armature voltage may be determined as

$$\bar{v}_a = \frac{1}{T}\left[\int_0^{t_1} v_S\, d\xi + \int_{t_1}^{T} 0\, d\xi\right] \tag{5.5-1}$$

Since $t_1 = kT$, where k is referred to as the duty cycle, the average armature voltage becomes

$$\bar{v}_a = kv_S \tag{5.5-2}$$

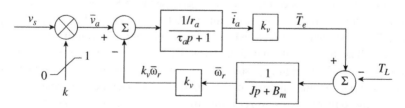

Figure 5.5-3 Average-value model of two-quadrant dc converter drive.

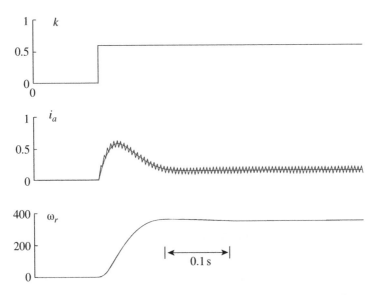

Figure 5.5-4 Starting characteristics of a permanent-magnet dc machine with a two-quadrant dc/dc converter drive.

In Fig. 5.5-3, the bars over the variables denote average values.

The starting characteristics of a permanent-magnet dc machine with a two-quadrant chopper drive are depicted in Fig. 5.5-4. The machine parameters are $r_a = 7\,\Omega$, $L_{AA} = 120\,\mathrm{mH}$, $k_v = 1.41 \times 10^{-2}\,\mathrm{V \cdot s/rad}$, and $J = 1.06 \times 10^{-6}\,\mathrm{kg \cdot m^2}$; rated voltage is 6 V. Here, the switching frequency f_s is set to 200 Hz and the source voltage to 10 V. Typically, the switching frequency is much higher, generally greater than 20 kHz. The frequency was selected to illustrate the dynamics introduced by the converter. Even at this low switching frequency, the switching period T is much less than the armature time constant τ_a. Thus, the armature current essentially consists of piecewise linear segments about an average response. In Fig. 5.5-4, the duty cycle is stepped from 0 to 0.6, corresponding to a step increase in average applied voltage from 0 to 6 V. The start-up response established using the average-value model is superimposed for purposes of comparison. As shown, the only salient difference between the two responses is the "sawtooth" behavior of the armature current due to converter switching. The difference in rotor speeds is indistinguishable.

5.5.2 Torque Control

The parameters of a permanent-magnet dc machine are $V_a = 6\,\mathrm{V}$ rated, $r_a = 7\,\Omega$, $k_v = 1.41 \times 10^{-2}\,\mathrm{V \cdot s/rad}$, $L_{AA} = 120\,\mathrm{mH}$, $J = 1.06\,\mathrm{kg \cdot m^2}$, and $B_m = 6.04 \times 10^{-6}\,\mathrm{N \cdot m \cdot s}$. We are to limit the torque T_e^* to $0.423 \times 10^{-2}\,\mathrm{N \cdot m}$ or

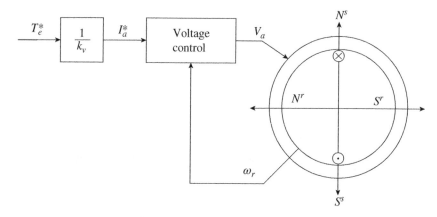

Figure 5.5-5 Torque control.

$I_a^* = 0.3$ A where the asterisk denotes commanded values. The control is shown symbolically in Fig. 5.5-5.

Since the current is controlled, the electric dynamics are neglected; therefore, only the mechanical dynamics are considered. The equations involved in Fig. 5.5-5 are

$$V_a = r_a I_a^* + \omega_r k_v \tag{5.5-3}$$

$$T_e^* = J \frac{d\omega_r}{dt} + B_m \omega_r + T_L \tag{5.5-4}$$

where $T_e^* = k_v I_a^* = 0.423 \times 10^{-2}$ N·m. The load line is

$$T_L = K \omega_r^2 \tag{5.5-5}$$

where

$$K = 5.529 \times 10^{-8} \text{ N·m·s}^2 \tag{5.5-6}$$

This intersects rated V_a torque versus rotor speed plot at Operating Point 1 where $\omega_r = 276.6$ rad/s, as shown in Fig. 5.5-6.

The dc machine is operating at point 1. The commanded torque is suddenly switched to $\frac{1}{2}$ the original value which intersects the limiting torque $\left(I_a^*\right)$ at Operating Point 2 where $\omega_r = 195.6$ rad/s. The electromechanical dynamics are

$$T_e^* = J \frac{d\omega_r}{dt} + B_m \omega_r + T_L \tag{5.5-7}$$

Assuming the torque control is functioning perfectly, the rotor slows and steady state is reached at Operating Point 2. The voltage at Operating Point 2 is

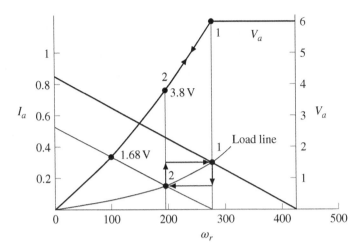

Figure 5.5-6 Drive operation during T_e^* switching.

$$V_a = r_a I_a^* + k_v \omega_r$$
$$= (7)(0.15) + (1.41 \times 10^{-2})(195.6) \qquad (5.5\text{-}8)$$
$$= 1.05 + 2.76 = 3.8 \text{ V}$$

The commanded torque T_e^* is returned to the original value. The rotor speeds up and reaches steady state at Operating Point 1. The trajectory from Operating Point 1 to Operating Point 2 and then back to 1 is shown in Fig. 5.5-6.

SP5.5-1. The dc machine given in the subsection entitled "Torque Control" is supplied from a two-quadrant converter. Determine k to reduce the no-load speed to one-half that when rated voltage is applied to the armature [$k = 0.3$].

SP5.5-2. Sketch Fig. 5.5-2 for only motor action; i.e. i_{S2} and i_{D1} equal to zero.

SP5.5-3. Assume $\omega_r = 100$ rad/s, calculate V_a and check the value with that given in Fig. 5.5-6 [1.68 V]

5.6 Problems

1 A permanent-magnet dc motor has the following parameters: $r_a = 8\ \Omega$ and $k_v = 0.01$ V·s/rad. The shaft load torque is approximated as $T_L = K\omega_r$, where $K = 5 \times 10^{-6}$ N·m·s. The applied voltage is 8 V and $B_m = 0$. Calculate the steady-state rotor speed ω_r in rad/s.

2 A permanent-magnet dc motor is driven by a mechanical source at 3820 r/min. The measured open-circuit armature voltage is 7 V. The mechanical source is disconnected, and a 12-V electric source is connected to the armature. With zero-load torque, $I_a = 0.1$ A and $\omega_r = 650$ rad/s. Calculate k_v, B_m, and r_a.

3 The parameters of a permanent-magnet dc machine are $r_a = 6\,\Omega$ and $k_v = 2 \times 10^{-2}\,\text{V} \cdot \text{s/rad}$. V_a can be varied from 0 to 10 V. The device is to be operated in the constant-torque mode with $T_e = 4 \times 10^{-3}\,\text{N} \cdot \text{m}$. (a) Determine V_a for $\omega_r = 0$. (b) Determine maximum ω_r range of the constant-torque mode of operation that is, maximum ω_r with $T_e = 4 \times 10^{-3}\,\text{N} \cdot \text{m}$ and $V_a = 10\,\text{V}$.

4 Sketch Fig. 5.5-2 for generator action; i.e. i_{S1} and i_{D2} equal to zero.

Reference

1 P. C. Krause, *Analysis of Electric Machinery*, New York, McGraw-Hill Book Company, 1986.

6

Brushless dc and Field-Oriented Drives

6.1 Introduction

The three-phase permanent-magnet ac machine supplied from a controlled inverter is used widely as a drive in low-to-medium power applications. Although it may seem as a contradiction of terms, this inverter and ac machine combination is often referred to as a brushless dc drive. Depending upon the control strategy and inverter used, the performance of this brushless dc drive can be made to (1) emulate the performance of a permanent-magnet dc motor, (2) operate in a maximum torque per volt mode, (3) or in a maximum torque per ampere mode. Fortunately, we are able to become quite familiar with the operating features of these various modes without becoming overly involved with the actual inverter control. In particular, if we assume that the stator variables (voltages and currents) are sinusoidal and balanced with the same angular velocity as the rotor speed, we are able to predict the dominant operating features of all modes of operation without becoming involved with the actual switching or control of the inverter. Therefore, we will focus on the performance of the brushless dc drive assuming that the inverter is designed and controlled appropriately. We will use the rotor reference frame to derive the equations used in the analysis and for the control of the machine.

Although this chapter is primarily devoted to the brushless dc drive, the last section deals with field orientation of an induction motor drive. The dominant characteristics of this drive are considered assuming the control is working perfectly.

Introduction to Modern Analysis of Electric Machines and Drives, First Edition.
Paul C. Krause and Thomas C. Krause.
© 2023 The Institute of Electrical and Electronics Engineers, Inc.
Published 2023 by John Wiley & Sons, Inc.

6.2 The Brushless dc Drive Configuration

The drive converter that we will consider is the three-phase six-step inverter shown in Fig 6.2-1 supplying the wye-connected symmetrical stator windings of a three-phase permanent-magnet ac machine shown in Fig. 6.2-2. The drive needs some way to measure the rotor speed of the machine, fortunately, there are many

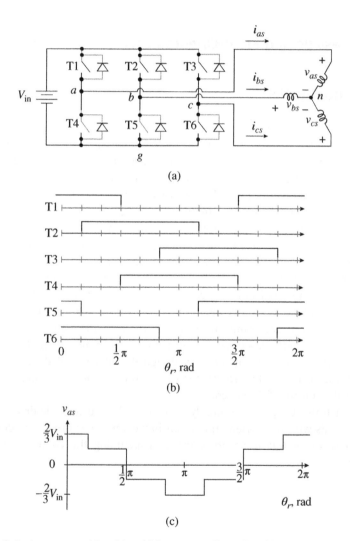

Figure 6.2-1 Inverter-machine drive. (a) Inverter configuration, (b) transistor switching logic, and (c) plot of v_{as}.

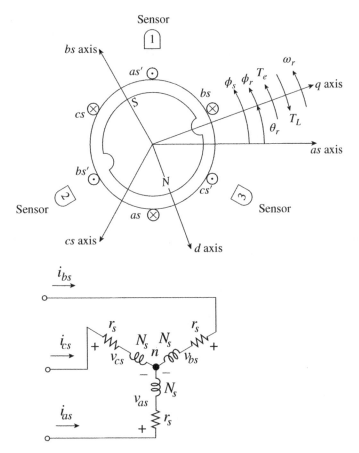

Figure 6.2-2 Two-pole three-phase permanent-magnet ac machine with sensors.

sensors and techniques to estimate rotor speed. For example, the three sensors shown in Fig. 6.2-2 are Hall-effect devices; when the south pole of the permanent-magnet rotor is under a sensor its output is nonzero; with the rotor north pole under the sensor, its output is zero. Regardless of the technique used, the rotor speed estimate determines the switching logic for the inverter, which, in turn, determines the output frequency and phase of the machine voltages. In the actual machine, the sensors are not positioned over the rotor as shown in Fig. 6.2-2. Instead, they are placed over a ring that is mounted on the shaft external to the stator and magnetized the same as the rotor.

The inverter shown in Fig. 6.2-1a consists of six transistors each with an antiparallel diode supplied from a dc source V_{in}. The logic (switching) signals for the transistors are shown in Fig. 6.2-1b. It is assumed that the forward resistance of the diodes and transistors is negligibly small and the turn-on and turn-off times are neglected, whereupon the transistors and diodes are considered to be ideal switches. With these assumptions, the instant one of the "top-rail" transistors in a phase is turned off the "bottom-rail" transistor of that phase can be turned on and conversely. In other words, each phase is either connected to the top or bottom rail; this is referred to as continuous-current or 180° mode of operation.

From the work in Example 1D, we can write

$$v_{ag} = v_{as} + v_{ng} \tag{6.2-1}$$

$$v_{bg} = v_{bs} + v_{ng} \tag{6.2-2}$$

$$v_{cg} = v_{cs} + v_{ng} \tag{6.2-3}$$

Adding (6.2-1) through (6.2-3) and since for a symmetrical wye-connected machine $v_{as} + v_{bs} + v_{cs} = 0$, we can write

$$v_{ng} = \frac{1}{3}\left(v_{ag} + v_{bg} + v_{cg}\right) \tag{6.2-4}$$

Substituting (6.2-4) into (6.2-1) through (6.2-3) and then solving for v_{as}, v_{bs}, and v_{cs} yields

$$v_{as} = \frac{2}{3}v_{ag} - \frac{1}{3}\left(v_{bg} + v_{cg}\right) \tag{6.2-5}$$

$$v_{bs} = \frac{2}{3}v_{bg} - \frac{1}{3}\left(v_{cg} + v_{ag}\right) \tag{6.2-6}$$

$$v_{cs} = \frac{2}{3}v_{cg} - \frac{1}{3}\left(v_{ag} + v_{bg}\right) \tag{6.2-7}$$

where v_{ag}, v_{bg}, and v_{cg} are either V_{in} or zero depending upon the state of the transistors. A plot of v_{as} is shown in Fig. 6.2-1c. The voltages v_{bs} and v_{cs} are of the same waveform as v_{as} but lag by 120° and 240°, respectively, for an *abc* sequence. The firing of the inverter is controlled so that the fundamental frequency of the applied voltages ($v_{as}, v_{bs},$ and v_{cs}) corresponds to the rotor speed ω_r of the permanent-magnet ac machine. That is, T1 is turned on by sensor 1, T2 by sensor 2, and T3 by sensor 3.

The plot of v_{as} given in Fig. 6.2-1c is readily established from (6.2-5) and the firing signals given in Fig. 6.2-1b. When T4, T5, and T6 are high terminals a, b, and c, respectively, are connected to the lower rail and are zero. They allow a return path for the currents. The signals T1, T2, and T3 determine the plot of v_{as}, v_{bs}, and v_{cs}. In the case of v_{as}, Fig. 6.2-1c tells us that when T1 is high and T2 and T3 are low, v_{as} is

$\frac{2}{3}V_{in}$. When T2 or T3 are high, they each contribute $-\frac{1}{3}V_{in}$ to v_{as}. Clearly v_{as} is $-\frac{2}{3}V_{in}$ when T1 is low and both T2 and T3 are high.

In order to analytically "connect" the inverter to the machine equations, we need to transform v_{as}, v_{bs}, and v_{cs} to v_{qs}^r and v_{ds}^r; the voltage equations used to analyze the permanent-magnet ac machine. A Fourier expansion of v_{as} shown in Fig. 6.2-1c may be written as

$$v_{as} = \frac{2V_{in}}{\pi}\left(\cos\theta_r + \frac{1}{5}\cos 5\theta_r - \frac{1}{7}\cos 7\theta_r + \cdots\right) \qquad (6.2\text{-}8)$$

The inverter controls the frequency of v_{as}, v_{bs}, and v_{cs} to correspond to ω_r, the electrical angular velocity of the rotor; hence, the voltages are expressed in terms of ω_r.

It should be noted that the maximum value of Fourier expansion of v_{as} given in Fig. 6.2-1c is $\frac{2}{3}V_{in}$; however, the peak value of the fundamental component of v_{as} given in (6.2-8) is $\frac{2}{\pi}V_{in}$. The voltages v_{bs} and v_{cs} may be expressed, for an abc sequence, by substituting $\theta_r - \frac{2}{3}\pi$ and $\theta_r + \frac{2}{3}\pi$, respectively, for θ_r in (6.2-8).

Let us take a closer look at the voltage and current waveforms of the continuous-current inverter. For this purpose, it is sufficient to focus on v_{as} and i_{as} shown in Fig. 6.2-3. We will work with Fig. 6.2-1a and Fig. 6.2-3, where v_{as}, i_{as}, i_{aT1}, $-i_{ad1}$, $-i_{aT4}$, and i_{ad4} are plotted for a typical operating condition with an RL load for $V_{in} = 25$ V, where $i_{ad1}(i_{ad4})$ is the current flowing in the antiparallel diode of T1(T4). The diodes are not labeled in Fig. 6.2-1a. Let us start at the center of the peak value of v_{as}. We see from Fig. 6.2-1a that at this instant T1 is carrying positive i_{as} and it is being returned to the source through T5(i_{bs}) and T6(i_{cs}) (not shown in Fig. 6.2-3). At the first step, T2 is turned on, T5 is turned off, and v_{as} drops to $\frac{1}{3}V_{in}$. Now, T1(i_{as}) and T2(i_{bs}) (not shown in Fig. 6.2-3) will share the positive current and T6(i_{cs}) will return the current to the source. At the next step, T1 is turned off and T4 is turned on. Now T2(i_{bs}) has all of positive current and T4(i_{as}) and T6(i_{cs}) will share the return current. However, since we have an inductive circuit, before T4 can share the return current there must be path for the positive current that was flowing in T1(i_{as}) at the time it was turned off. This is through i_{d4} (the antiparallel diode of T4). When T4 is turned off, the antiparallel diode of T1 springs into action. The pattern is clear.

If we use (2.3-18) with $\theta = \theta_r$ to transform v_{as}, v_{bs}, and v_{cs}, where v_{as} is given by (6.2-8), to the ω_r reference frame, we obtain

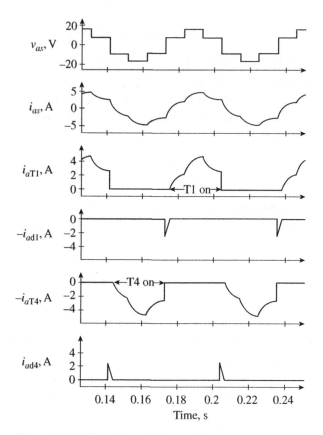

Figure 6.2-3 Plots of v_{as} and i_{as} and the components of i_{as}.

$$v_{qs}^r = \frac{2V_{in}}{\pi}\left(1 + \frac{2}{35}\cos 6\theta_r - \frac{2}{143}\cos 12\theta_r + \cdots\right) \tag{6.2-9}$$

$$v_{ds}^r = \frac{2V_{in}}{\pi}\left(\frac{12}{35}\sin 6\theta_r - \frac{24}{143}\sin 12\theta_r + \cdots\right) \tag{6.2-10}$$

The frequency of the harmonics has changed. The 5th and 7th of the three-phase voltages have become the 6th harmonic and the 11th and 13th have become the 12th in the q_s^r- and d_s^r-voltages. The balanced set of 5th and 11th harmonics produce magnetic fields that rotate cw at $5\omega_r$ and $11\omega_r$. The balanced 7th and 13th harmonics produce magnetic fields that rotate ccw at $7\omega_r$ and $13\omega_r$. When we transform ccw to ω_r, the 5th and 7th become the 6th and the 11th and 13th become the 12th.

SP6.2-1. Show that the 5th harmonic causes a rotating mmf in the cw direction at $5\omega_e$ in the steady state. [5th is an *acb* sequence]

SP6.2-2. Show that 7th causes a ccw rotation. [7th is an *abc* sequence]

6.3 Normal Mode of Brushless dc Drive Operation

The free-acceleration characteristics of this brushless dc motor-inverter drive are shown in Fig. 6.3-1. The three-phase four-pole machine parameters are $r_s = 3.4\,\Omega$, $L_{ls} = 1.1\,\text{mH}$, $L_{Ms} = \dfrac{3}{2}11\,\text{mH}$, $L_{ss} = 17.6\,\text{mH}$, and $\lambda''_m = 0.0826\,\text{V} \cdot \text{s/rad}$ with a total inertia of $5\times10^4\,\text{kg} \cdot \text{m}^2$. It is convenient to relate the phase of the three-phase applied voltages to the q axis as

$$\phi_v = \theta_{esv} - \theta_r \qquad (6.3\text{-}1)$$

Since $\omega_r = \omega_e$, steady-state operation (6.3-1) becomes

$$\phi_v = \theta_{esv}(0) - \theta_r(0) \qquad (6.3\text{-}2)$$

Also, for brushless dc machine operation, it is convenient to select $\theta_r(0)$ equal to zero, whereupon

$$\phi_v = \theta_{esv}(0) \qquad (6.3\text{-}3)$$

Therefore, ϕ_v is the phase of $\tilde{V}_{as}$ and the phase of $\tilde{E}_a$ is zero degrees since it is along the q axis.

The normal mode of operation of the brushless dc drive is to control the inverter so that the middle of the maximum v_{as} coincides with the q-axis, $\phi_v = 0$, with the frequency the three-phase voltages equal to the electrical angular velocity of the rotor. The free-acceleration characteristics of this brushless dc motor-inverter drive with $\phi_v = 0$ are shown in Fig. 6.3-1. The T_e versus ω_r for Fig. 6.3-1 is shown in Fig. 6.3-2. Note the similarity of the steady-state torque given in Fig. 6.3-2 and that given for the permanent magnet dc motor given in Fig. 5.4-1. Thus, the name brushless dc.

The free-acceleration characteristics shown in Figs. 6.3-3 and 6.3-4 are with the harmonics neglected in (6.2-9) and (6.2-10). Neglecting the voltage harmonics is a good approximation except for the minor difference since the magnitude of v_{as} is $\dfrac{2}{3}V_{in}$ in Figs. 6.3-1 and 6.3-2 and $\dfrac{2}{\pi}V_{in}$ in Figs. 6.3-3 and 6.3-4. We will neglect the voltage harmonics in future work.

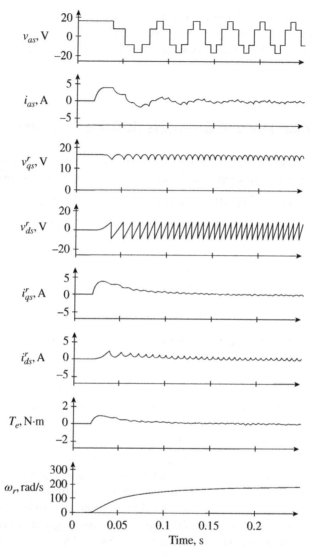

Figure 6.3-1 Free-acceleration characteristics of a three-phase brushless dc motor supplied by a six-step inverter with $\phi_v = 0$.

In Fig. 6.3-5, the load torque is initially 0.1 N · m and the machine is operating in the steady state. The load torque is stepped to 0.4 N · m, the machine slows and steady-state operation is established with $T_L = 0.4$ N·m, whereupon the load torque is stepped back to 0.1 N · m. Note the change in frequency of that stator voltage

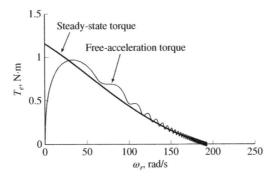

Figure 6.3-2 Torque-speed characteristics for free acceleration shown in Fig 6.4-1.

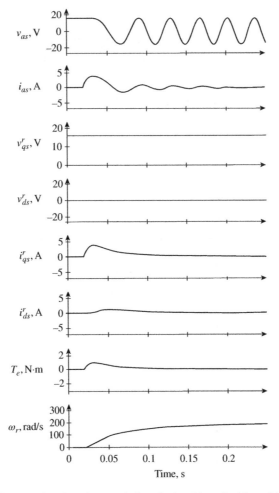

Figure 6.3-3 Free-acceleration characteristics of a brushless dc drive with harmonics neglected.

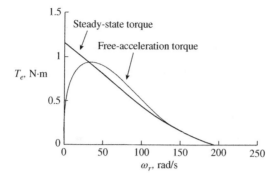

Figure 6.3-4 Torque-speed characteristics for the free acceleration shown in Fig. 6.3-3 with the steady-state torque shown for comparison purposes.

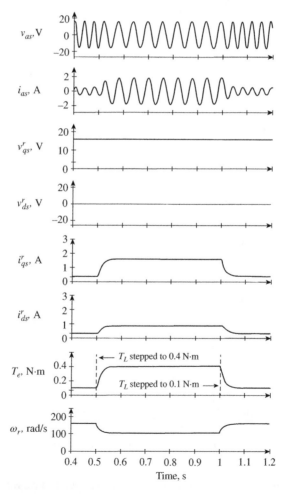

Figure 6.3-5 Dynamic performance of a brushless dc drive during step changes in load torque. Total inertia of 2×10^{-4} kg·m^2.

and current and also the increase in stator current due to the decrease in inductive reactances until $T_e = T_L$.

Before leaving this section, let us derive the instantaneous and steady-state phasor voltage equations by following the work in Section 2.4. In the case of the permanent-magnet ac machine, the electrical angular velocity (frequency) of the stator applied voltages is controlled by the inverter to be the same as the electrical angular velocity of the rotor; in other words ω_e is controlled to always be ω_r. Therefore, as long as $\omega_e = \omega_r$, the instantaneous and steady-state phasors may be used to portray transient and steady-state machine and drive operation. If the rotor and stator are symmetrical as in the case of an induction machine, all balanced modes of operation regardless of rotor speed yield constant steady-state synchronous reference frame variables. In the case of the permanent-magnet ac machine, which is an unsymmetrical machine due to the rotor, yields constant synchronous reference frame variables in the steady state only if the rotor speed is fixed at synchronous speed. This is not the case in general; however, we will find that when the permanent magnet ac machine is controlled as a brushless dc machine, the fundamental frequency of the applied stator voltages is controlled essentially, instantaneously to be equal to the rotor speed. Therefore, all modes of operation are in the synchronous reference frame and the phasor representation is valid if we neglect the harmonics introduced in the stator voltages due to the switching of the drive inverter.

Substituting (4.2-4) and (4.2-5) into (2.4-32) with r used as a superscript rather than e since $\omega_r = \omega_e$, we have

$$\left(v_{qs}^r - jv_{ds}^r\right) = r_s\left(i_{qs}^r - ji_{ds}^r\right) + L_{ss}\left(pi_{qs}^r - jpi_{ds}^r\right)$$
$$+ \omega_r L_{ss}\left(i_{ds}^r + ji_{qs}^r\right) + \omega_r \lambda_m' \tag{6.3-4}$$

which may be written as

$$\tilde{v}_{as} = (r_s + j\omega_r L_{ss})\tilde{i}_{as} + \tilde{e}_a + L_{ss}p\tilde{i}_{as} \tag{6.3-5}$$

where we are assuming that

$$\tilde{e}_a = \omega_r \lambda_m''^{r}\underline{/0°} \tag{6.3-6}$$

which is added to (2.4-37) due to the rotor.

Now, for balanced steady-state operation, the last term of (6.3-5) is zero and the steady-state phasor voltage equation becomes

$$\tilde{V}_{as} = (r_s + j\omega_r L_{ss})\tilde{I}_{as} + \tilde{E}_a \tag{6.3-7}$$

where

$$\tilde{E}_a = \frac{\omega_r \lambda_m''^{r}}{\sqrt{2}}\underline{/0°} \tag{6.3-8}$$

For a three-phase machine, we will replace L_{ms}, which is part of L_{ss}, with L_{Ms} or $\frac{3}{2}L_{ms}$ in (6.3-5) and (6.3-7).

We have set $\theta(0) = 0$, therefore the q-axis is fixed at zero degrees, thus since $\omega_r = \omega_e$ we will also set $\theta_r(0) = 0$. Therefore, since $\phi_v = \theta_{esv}(0)$ $\tilde{V}_{as}$, and $\tilde{E}_a$ are in the q-axis. A convenient expression for the calculation of steady-state torque for balanced operation can be obtained from (4.2-28) by expressing the instantaneous three-phase currents as given by (2.2-15) through (2.2-17) and transforming the currents by (2.3-18) to the rotor reference. Substituting I^r_{qs} into (4.2-28) gives the torque as

$$T_e = \frac{P}{2}\frac{3}{2}\lambda'^r_m\sqrt{2}I_s\cos[\theta_{esi}(0) - \theta_r(0)] \tag{6.3-9}$$

We can use the phasor diagram to advantage in portraying the steady-state operation of the permanent-magnet ac machine. During balanced steady-state operation and with the assumption of $\omega_e = \omega_r$, the rotor and phasor diagram are rotating in unison; therefore, they can be superimposed with $\tilde{E}_a$ at zero degrees. Moreover, since mmf$_r$ is also rotating at ω_r, N^r and S^r can be superimposed with the rotor and the phasor diagram. Now, since all are rotating in unison and the voltage and current phasors are constant, we can stop the rotation at any time or we can run at ω_e and observe the operation of the machine.

It is important to realize that we have positioned the rotor so that when a voltage is applied at stall, the torque is nearly maximum in the ccw direction. Depending on the position of the rotor at stall, the starting torque may be small or even in the cw direction. The control must take this into account to prevent this. Also, when the torque is in ccw direction and the rotor speed is increasing as shown in Figs. 6.3-1 through 6.3-4, the speed at which the torque becomes zero depends upon the relative value of the applied voltage, and λ'_m. That is, with $\phi_v = 0$, when $\tilde{V}_{as} = \tilde{E}_a$ the stator current and thus the torque become zero [see (6.3-7) or (6A-2)].

Example 6.A   Steady-State Operation with $\phi_v = 0$

Let us assume that the parameters of the machine are those given earlier in this section with $V_{in} = 25$ V and the actual steady-state rotor speed is 750 r/min. Calculate ω_e, T_e, and draw the phasor diagram showing the stator and rotor poles.

From (3.6-9), the electrical angular velocity of the rotor is

$$\begin{aligned}
\omega_r &= \frac{P}{2}\omega_{rm}\\
&= \frac{P}{2}\frac{(\text{r/min})(\text{rad/r})}{\text{s/min}}\\
&= \frac{4}{2}\frac{(750)(2\pi)}{60} = 50\pi \text{ rad/s}
\end{aligned} \tag{6A-1}$$

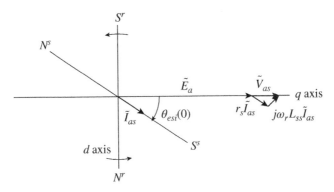

Figure 6.A-1   Phasor diagram for operation at $\omega_r = 50\pi$ rad/s with $\phi_v = 0$.

Using phase variables, we can solve for $\tilde{I}_{as}$ from (6.3-7) with $L_{ss} = L_{ls} + L_{Ms}$,

$$
\begin{aligned}
\tilde{I}_{as} &= \frac{\tilde{V}_{as} - \tilde{E}_a}{r_s + j\omega_r L_{ss}} \\
&= \frac{11.25\underline{/0^\circ} - 9.18\underline{/0^\circ}}{3.4 + j50\pi(17.6 \times 10^{-3})} \\
&= \frac{2.07\underline{/0^\circ}}{3.4 + j2.76} = \frac{2.07\underline{/0^\circ}}{4.38\underline{/39.1^\circ}} = \underline{0.473\underline{/-39.1^\circ}\text{ A}}
\end{aligned} \tag{6A-2}
$$

The torque can be calculated using (6.3-9), thus,

$$
\begin{aligned}
T_e &= \frac{P}{2}\left(\frac{3}{2}\right)\lambda'_m\sqrt{2}I_s\cos[\theta_{esi}(0) - \theta_r(0)] \\
&= \frac{4}{2}\frac{3}{2}(0.0826)\sqrt{2}(0.473)\cos(-39.1^\circ - 0) \\
&= 0.129 \text{ N·m}
\end{aligned} \tag{6A-3}
$$

The phasor diagram is shown in Fig. 6.A-1.

We see from Fig. 6.A-1 that the rotor poles are being "pulled" in the counter-clockwise direction by the poles created by the stator currents, motor action.

SP6.3-1. Plot the trajectory of the torque switching of Fig. 6.3-5 on the steady-state torque plot shown in Fig. 6.3-4.

SP6.3-2. If $\omega_r = \omega_e$ and $\theta_r(0) = 0$ show that $v^r_{ds} = 0$ for $\phi_v = 0$. $[\theta_{esv}(0) = 0]$

SP6.3-3. With the initial conditions assumed, the stall torque of an inverter-driven brushless dc machine is slightly larger than the sinusoidal approximation. Why? $\left[\frac{2}{3}V_{in} \text{ vs } \frac{2}{\pi}V_{in}\right]$

SP6.3-4. For a given inverter voltage what factors determine the starting torque. [rotor position, position of mmf$_s$]

6.4 Other Modes of Brushless dc Drive Operation

There are two other modes of ϕ_v that we will consider: the maximum torque per voltage, $\phi_{vMT/V}$, and the maximum torque per ampere, $\phi_{vMT/A}$. The final mode of operation that we will consider is the torque control mode which requires a reduction in the applied voltage as speed decreases. Although $\phi_v = 0$ is a common mode of operation of the brushless dc drive, researchers in [1, 2] discovered that advancing ϕ_v with respect to the q axis could increase the torque at rotor speeds greater than zero. This was shown analytically in [3] and illustrated by simulating the phase shifting (increasing ϕ_v) of the applied voltages to obtain maximum torque per volt $\left(\phi_v = \phi_{vMT/V} \right)$ at a given speed.

If the applied voltages and thus the stator poles are shifted relative to the magnetic field established by the permanent-magnet rotor, which is fixed in the d axis, the torque versus speed characteristics can be changed over a wide range by shifting ϕ_v from zero to 2π [4]. Here, we limit our discussion to shifting ϕ_v for the purpose of maximizing torque during motor operation.

6.4.1 Maximum-Torque Per Volt Operation of a Brushless dc Drive $(\phi_v = \phi_{vMT/V})$

Torque is proportional to i_{qs}^r and when ϕ_v is shifted from zero, v_{ds}^r is nonzero. For the purpose of deriving an expression for the maximum torque per volt at a given rotor speed $\left(\phi_{vMT/V} \right)$, we will start with the steady-state versions of (4.2-4) and (4.2-5) and (4.2-21) and (4.2-22) for V_{qs}^r and V_{ds}^r, respectively. In particular, since $\omega_r = \omega_e$, the derivative terms are zero for steady-state operation. Thus,

$$V_{qs}^r = r_s I_{qs}^r + \omega_r L_{ss} I_{ds}^r + \omega_r \lambda_m^{\prime r} \tag{6.4-1}$$

$$V_{ds}^r = r_s I_{ds}^r - \omega_r L_{ss} I_{qs}^r \tag{6.4-2}$$

Please recall that these equations are valid for two- or three-phase devices for balanced steady-state operation if, for the three-phase device, L_{ms} is replaced by $\frac{3}{2} L_{ms}$ is used when calculating L_{ss}. We also need the expressions for V_{qs}^r and V_{ds}^r as functions of ϕ_v We can obtain these relationships by transforming V_{as} and V_{bs} for the two-phase machine or V_{as}, V_{bs}, and V_{cs} for the three-phase, to the rotor reference frame and substituting (6.3-1) for ϕ_v. That is,

$$V_{qs}^r = \sqrt{2} V_s \cos \phi_v \tag{6.4-3}$$

$$V_{ds}^r = -\sqrt{2} V_s \sin \phi_v \tag{6.4-4}$$

Since $\theta_r(0) = 0$, ϕ_v is the phase of $\tilde{V}_{as}$.

Solving (6.4-2) for I_{ds}^r and substituting the result into (6.4-1) yields

$$V_{qs}^r = \frac{r_s^2 + \omega_r^2 L_{ss}^2}{r_s} I_{qs}^r + \frac{\omega_r L_{ss}}{r_s} V_{ds}^r + \omega_r \lambda_m''^r \qquad (6.4\text{-}5)$$

Now, solving (6.4-5) for I_{qs}^r and substituting (6.4-3) and (6.4-4) for V_{qs}^r and V_{ds}^r, respectively, with $\theta_r(0) = 0$, we have

$$I_{qs}^r = \frac{r_s}{r_s^2 + \omega_r^2 L_{ss}^2} \left(\sqrt{2} V_s \cos \phi_v + \frac{\omega_r L_{ss}}{r_s} \sqrt{2} V_s \sin \phi_v - \omega_r \lambda_m''^r \right) \qquad (6.4\text{-}6)$$

It is interesting to note from (6.4-6) that V_{ds}^r aids V_{qs}^r to increase I_{qs}^r for a given rotor speed. Since this results in a negative I_{ds}^r, it is often referred to as field weakening even though $\lambda_m''^r$ is not decreased in magnitude.

Since T_e is proportional to I_{qs}^r, (4.2-28), we can obtain the maximum torque for a given rotor speed by taking the derivative of I_{qs}^r with respect to ϕ_v and setting the result equal to zero and then solving for ϕ_v. Thus, from (6.4-6),

$$0 = -\sin \phi_v + \frac{\omega_r L_{ss}}{r_s} \cos \phi_v \qquad (6.4\text{-}7)$$

whereupon

$$\frac{\sin \phi_v}{\cos \phi_v} = \frac{\omega_r L_{ss}}{r_s} \qquad (6.4\text{-}8)$$

or

$$\phi_{vMT/V} = \tan^{-1} \frac{\omega_r L_{ss}}{r_s} \qquad (6.4\text{-}9)$$

Equation (6.4-9) tells us that for a given positive rotor speed, $\phi_{vMT/V}$ will yield maximum possible torque per volt at that rotor speed. Equation (6.4-9) is derived for steady-state conditions and is in slight error for transient conditions.

The free-acceleration characteristics for $\phi_{vMT/V}$ are shown in Figs. 6.4-1 and 6.4-2. These characteristics may be compared to Figs. 6.3-3 and 6.3-4, respectively, where $\phi_v = 0$. Note the extended speed range with $\phi_{vMT/V}$ (Fig. 6.4-2) compared with $\phi_v = 0$ (Fig. 6.3-4). Also, note that i_{ds}^r is small positive in Fig. 6.3-3 but a larger negative value in Fig. 6.4-1. In other words, an increase in torque $\left(I_{qs}^r \right)$ and speed range occurs due to a decrease in I_{ds}^r (a larger negative value).

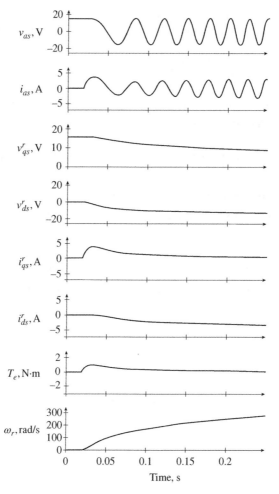

Figure 6.4-1 Free-acceleration characteristics of a brushless dc drive with $\phi_v = \phi_{vMT/V}$ and a total inertia of 5×10^{-4} kg · m².

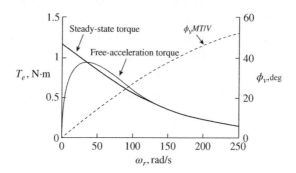

Figure 6.4-2 Torque-speed characteristics for free acceleration shown in Fig. 6.4-1 with the steady-state torque also shown. Compare to Fig. 6.3-4 where $\phi_v = 0$.

Example 6.B   Brushless dc Drive Operation with $\phi_v = \phi_{vMT/V}$

In Example 6.A, we determined the steady-state torque at 750 r/min ($\omega_r = 50\pi$ rad/s) with $\phi_v = 0$. In this example, we will calculate the torque at the same speed with $\phi_v = \phi_{vMT/V}$, draw the phasor diagram and locate the poles, and calculate the power balance. The parameters for the machine are repeated here for convenience; $r_s = 3.4\,\Omega$, $L_{ls} = 1.1$ mH, $L_{ms} = 11$ mH, and $\lambda''_m = 0.0826$ V·s/rad. From (6A-1), $\omega_r = 50\pi$ rad/s. Now, for a three-phase machine,

$$
\begin{aligned}
L_{ss} &= L_{ls} + \frac{3}{2}L_{ms} \\
&= \left(1.1 + \frac{3}{2}11\right) \times 10^{-3} = 17.6 \text{ mH}
\end{aligned}
\tag{6B-1}
$$

From (6.4-9),

$$
\begin{aligned}
\phi_{vMT/V} &= \tan^{-1}\frac{\omega_r L_{ss}}{r_s} \\
&= \tan^{-1}\frac{50\pi(17.6) \times 10^{-3}}{3.4} = 39.1°
\end{aligned}
\tag{6B-2}
$$

Therefore,

$$
V_{as} = \sqrt{2}11.25 \cos(\omega_r t + 39.1°)
\tag{6B-3}
$$

From (6.3-7),

$$
\tilde{V}_{as} = (r_s + j\omega_r L_{ss})\tilde{I}_{as} + \tilde{E}_a
\tag{6B-4}
$$

and from Example 6.A, $\tilde{E}_a = 9.18\underline{/0°}$. Solving for $\tilde{I}_{as}$,

$$
\begin{aligned}
\tilde{I}_{as} &= \frac{\tilde{V}_{as} - \tilde{E}_a}{r_s + j\omega_r L_{ss}} \\
&= \frac{11.25\underline{/39.1°} - 9.18\underline{/0°}}{3.4 + j50\pi(17.6) \times 10^{-3}} = 1.62\underline{/54.5°} \text{ A}
\end{aligned}
\tag{6B-5}
$$

From (6.3-9) the steady-state torque is

$$
\begin{aligned}
T_e &= \frac{3}{2}\frac{P}{2}\lambda''_m\sqrt{2}I_s \cos\theta_{esi}(0) \\
&= \left(\frac{3}{2}\right)\left(\frac{4}{2}\right)(0.0826)\left(\sqrt{2}\,1.62\cos 54.5°\right) = 0.330 \text{ N·m}
\end{aligned}
\tag{6B-6}
$$

The phasor diagram is shown in Fig. 6.B-1.

Calculating a power balance

$$
\begin{aligned}
P_e &= 3|\tilde{V}_{as}||\tilde{I}_{as}| \cos[\theta_{esv}(0) - \theta_{esi}(0)] \\
&= 3 \times 11.25 \times 1.62 \times \cos(39.1° - 54.5°) = 52.71 \text{ W}
\end{aligned}
\tag{6B-7}
$$

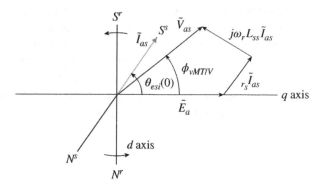

Figure 6.B-1 Phasor diagram for brushless dc drive operation at $\omega_r = 50\pi$ rad/s with $\phi_v = \phi_{vMT/V}$.

$$P_{loss} = 3r_s|\tilde{I}_{as}|^2$$
$$= 3 \times 3.4 \times (1.62)^2 = 26.77 \text{ W} \tag{6B-8}$$

$$P_m = T_e\omega_{rm}$$
$$= 0.33 \times \frac{2}{4} \times 50\pi = 25.92 \text{ W} \tag{6B-9}$$

The efficiency is

$$\text{eff} = \frac{P_m}{P_e} = \frac{25.92}{52.71} = 49.2\% \tag{6B-10}$$

From Fig. 6.B-1, we see that the poles created by the stator can be considered as "pushing" the rotor poles in the counterclockwise direction. In Example 6.A, T_e was calculated to be 0.129 N · m with $\phi_v = 0$. Here, with $\phi_{vMT/V}$, the torque is calculated to be 0.330 N · m; however, the magnitude of $\tilde{I}_{as}$ increased from 0.475 to 1.62 A. The torque is increased by a factor of 2.6 while the current is increased by a factor of 3.4. This is the maximum torque per volt that this device can produce at $\omega_r = 50\pi$ rad/s (750 r/min) with $V_s = 11.25$ V; however, when shifting the phase of the terminal voltage to achieve maximum torque, one must not exceed rated conditions for an extended period.

6.4.2 Maximum-Torque Per Ampere Operation of a Brushless dc Drive ($\phi_v = \phi_{vMT/A}$)

Maximum-torque per ampere operation occurs when I_{ds}^r (imaginary part of $\tilde{I}_{as}$) is made zero by controlling the position of $\tilde{V}_{as}$ relative to the permanent magnet of the rotor $\left(\phi_v = \phi_{vMT/A}\right)$. This was done by M. Hasan in [3]. The torque is directly related to the q-axis current (real part of $\tilde{I}_{as}$). The d-axis current does contribute to the torque indirectly, but decreases the efficiency of the machine as we have seen in Example 6.B.

To derive an expression for $\phi_{vMT/A}$ for steady-state operation, we will substitute (6.4-3) and (6.4-4) into (6.4-1) and (6.4-2) for V_{qs}^r and V_{ds}^r, respectively, and solve for $\cos\phi_v$ and $\sin\phi_v$. If we set $I_{ds}^r = 0$ and perform several mathematical manipulations, we can express $\phi_{vMT/A}$, at a given rotor speed, as

$$\phi_{vMT/A} = \tan^{-1}\left[\omega_r\tau_s\left(\frac{-1 \pm \omega_r\tau_v\sqrt{1 + \omega_r^2\tau_v^2\left(1 - \omega_r^2\tau_v^2\right)}}{\omega_r^4\tau_s^2\tau_v^2 - 1}\right)\right] \tag{6.4-10}$$

where, for compactness

$$\tau_s = \frac{L_{ss}}{r_s} \tag{6.4-11}$$

$$\tau_v = \frac{\lambda_m^{\prime r}}{\sqrt{2}V_s} \tag{6.4-12}$$

The free-acceleration characteristics are shown in Figs. 6.4-3 and 6.4-4 for $\phi_{vMT/A}$.

Example 6.C Brushless dc Drive Operation with $\phi_v = \phi_{vMT/A}$

In this example, we will calculate the torque with $\phi_v = \phi_{vMT/A}$ and draw the phasor diagram and locate the stator and rotor poles. The parameters are repeated again: $r_s = 3.4\,\Omega$, $L_{ls} = 1.1$ mH, $L_{ms} = 11$ mH, and $\lambda_m^{\prime r} = 0.0826$ V·s/rad. In order to compare with Examples 6.A and 6.B, we will perform the calculations for $\omega_r = 50\pi$ rad/s. From (6.4-10), $\phi_{vMT/A}$ is calculated to be 8.15° and 17° at $\omega_r = 50\pi$. The larger angle (17°) is an extraneous root. Also, $V_s = 11.25$ V, $L_{ss} = 17.6$ mH, and

$$V_{as} = \sqrt{2}\,11.25\cos\left(\omega_r t + 8.15°\right) \text{ V} \tag{6C-1}$$

From (6.3-7),

$$\tilde{V}_{as} = (r_s + j\omega_r L_{ss})\tilde{I}_{as} + \tilde{E}_a \tag{6C-2}$$

and again $\tilde{E}_a = 9.18\underline{/0°}$. Solving for $\tilde{I}_{as}$,

$$\tilde{I}_{as} = \frac{\tilde{V}_{as} - \tilde{E}_a}{r_s + j\omega_r L_{ss}}$$

$$= \frac{11.25\underline{/8.15°} - 9.18\underline{/0°}}{3.4 + j50\pi(17.6) \times 10^{-3}} = 0.577\underline{/0°} \text{ A} \tag{6C-3}$$

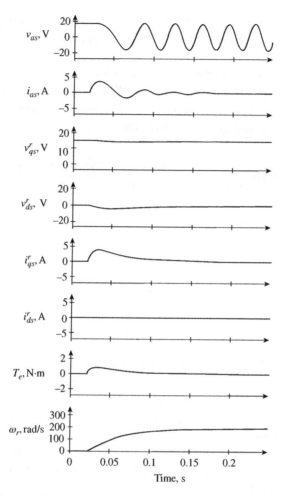

Figure 6.4-3 Free-acceleration characteristics of a brushless dc drive with $\phi_v = \phi_{vMT/A}$ $(i_{ds}^r = 0)$ and a total inertia of 5×10^{-4} kg·m².

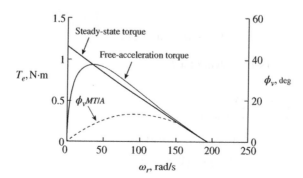

Figure 6.4-4 Torque-speed characteristics for free-acceleration shown in Fig. 6.4-3.

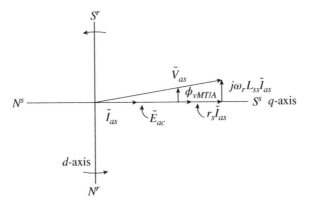

Figure 6.C-1 Phasor diagram for brushless dc drive operation at $\omega_r = 50\pi$ rad/s with $\phi_v = \phi_{vMT/A}$.

From (6.3-9),

$$
\begin{aligned}
T_e &= \frac{3}{2}\frac{P}{2}\lambda'^r_m\sqrt{2}I_s\cos\theta_{esi}(0) \\
&= \left(\frac{3}{2}\right)\left(\frac{4}{2}\right)(0.0826)\left(\sqrt{2}\right)(0.577) = 0.202 \text{ N·m}
\end{aligned}
\tag{6C-4}
$$

The phasor diagram is shown in Fig. 6.C-1. Note that the stator and rotor poles are orthogonal which yields the maximum torque per ampere for this device at $\omega_r = 50\pi$ rad/s.

The power balance becomes

$$
\begin{aligned}
P_e &= 3\left|\tilde{V}_{as}\right|\left|\tilde{I}_{as}\right|\cos\left[\theta_{esv}(0)-\theta_{esi}(0)\right] \\
&= 3 \times 11.25 \times 0.577\cos\left(8.15° - 0\right) = 19.3 \text{ W}
\end{aligned}
\tag{6C-5}
$$

$$
\begin{aligned}
P_{loss} &= 3r_s\left|\tilde{I}_{as}\right|^2 \\
&= 3 \times 3.4(0.577)^2 = 3.4 \text{ W}
\end{aligned}
\tag{6C-6}
$$

$$
\begin{aligned}
P_m &= T_e\omega_{rm} \\
&= 0.202 \times \frac{2}{4} \times 50\pi = 15.9 \text{ W}
\end{aligned}
\tag{6C-7}
$$

The efficiency is

$$
\text{eff} = \frac{P_m}{P_e} = \frac{15.9}{19.3} = 82.4\%
\tag{6C-8}
$$

6.4.3 Torque Control

In order to control the torque over a speed range, it is necessary to reduce the voltage applied to the machine. This can be done with pulse width modulation (PWM) of the drive inverter. This is accomplished by periodically connecting all three terminals of the six-step inverter to the bottom rail. The torque speed characteristics for reduced rms stator voltages are shown for $\phi_v = 0$ in Fig. 6.4-5.

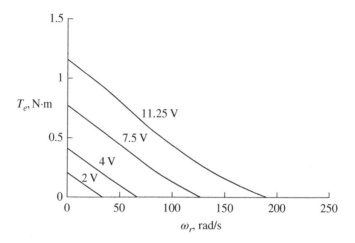

Figure 6.4-5 Torque-speed characteristics for $\phi_v = 0$ for different rms values of the applied voltage.

Example 6.D Torque Control

The parameters of the four-pole three-phase permanent magnet ac machine are $r_s = 3.4\,\Omega$, $L_{ls} = 1.1\,\text{mH}$, $L_{Ms} = 16.5\,\text{mH}$, $\lambda'_m = 0.0826\,\text{V}\cdot\text{s/rad}$, and rated $V_s = 11.25\,\text{V}$. The torque is controlled with $\phi_v = 0$ at $0.315\,\text{N}\cdot\text{m}$ until rated voltage is reached at $\omega_{rm} = 120.1\,\text{rad/s}$. The voltage, V_s, is then maintained at $11.25\,\text{V}$ with $\phi_v = 0$ and the load line and Operating Point 1 are as shown in Fig. 6.D-1. The load line is calculated from

$$T_L = K\omega_{rm}^2 \tag{6D-1}$$

where $T_L = 0.315\,\text{N}\cdot\text{m}$ and $\omega_r = 120.1\,\text{rad/s}$. Substituting into (6D-1) and solving for K yields

$$K = \frac{0.315}{(120.1)^2} = 2.184 \times 10^{-5}\,\text{N}\cdot\text{m}\cdot\text{s} \tag{6D-2}$$

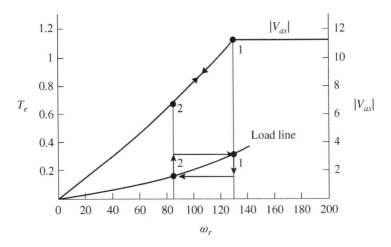

Figure 6.D-1 Controlling T_e with $\phi_v = 0$.

The commanded torque T_e^* is suddenly reduced to one half the original. Thus,

$$T_e^* = \left(\frac{1}{2}\right)(0.315) \tag{6D-3}$$
$$= 0.1575 \text{ N} \cdot \text{m}$$

The trajectory from Operating Point 1 to Operating Point 2 is shown in Fig. 6.D-1. The rotor speed at Operating Point 2 can be calculated as

$$T_L = K\omega_r^2 \tag{6D-4}$$

where $T_L = 0.1575$ N $\cdot$ m and $K = 2.184 \times 10^{-5}$ N $\cdot$ m $\cdot$ s^2. This yields

$$\omega_r = \left(\frac{0.1575}{2.184 \times 10^{-5}}\right)^{1/2} = 84.9 \text{ rad/s} \tag{6D-5}$$

Now, from (6.3-9),

$$T_e = \left(\frac{3}{2}\right)\left(\frac{P}{2}\right)\lambda''_m \sqrt{2}I_s \cos\theta_{esi}(0) \tag{6D-6}$$

We can write

$$\sqrt{2}I_s \cos\theta_{esi}(0) = \frac{T_e}{(3/2)(P/2)\lambda''_m} = \frac{0.1575}{(3/2)(4/2)0.0826} = 0.636 \text{ A} \tag{6D-7}$$

We can calculate the voltage from (6.3-7) and (6.3-8); however, we first need $\tilde{I}_{as}$. Since $\phi_v = 0$, $\tilde{V}_{as}$, and $\tilde{E}_a$ are at zero degrees, the phase angle of $\tilde{I}_{as}$ is

$$\theta_{est}(0) = - \tan^{-1} \frac{\omega_r L_{ss}}{r_s}$$

$$= - \tan^{-1} \frac{(84.9)\left(1.1 + \frac{3}{2} 11\right) \times 10^{-3}}{r_s} \tag{6D-8}$$

$$= - \tan^{-1} 0.439 = -23.7°$$

From (6D-7),

$$I_s = \frac{0.636}{\sqrt{2} \cos - 23.7°} = 0.4912 \text{ A} \tag{6D-9}$$

Therefore,

$$\tilde{I}_{as} = 0.4912 \underline{/-23.7°} \text{ A} \tag{6D-10}$$

Substituting into (6.3-7) and (6.3-8) yields

$$\tilde{V}_{as} = (r_s + j\omega_r L_{ss})\tilde{I}_{as} + \frac{1}{\sqrt{2}}\omega_r \lambda''_m \underline{/0°}$$

$$= (3.4 + j84.9)\left(1.1 + \frac{3}{2}11 \times 10^{-3}\right)\left(0.4912\underline{/-23.7°}\right) + \frac{(84.9)(0.0826)}{\sqrt{2}}$$

$$= \left(3.714\underline{/27.3°}\right)\left(0.4912\underline{/-23.7°}\right) + 4.96 = 6.78 \text{ V}$$

$$\tag{6D-11}$$

After steady-state operation is reached at Operating Point 2, the commanded torque is increased back to 0.315 N · m. The trajectory from Operating Point 2 to Operating Point 1 is shown in Fig. 6.D-1. The mechanical dynamics are determined by

$$T_e^* = J\left(\frac{2}{P}\right)\frac{d\omega_r}{dt} + B_m\left(\frac{2}{P}\right)\omega_r + T_L \tag{6D-12}$$

where $J = 5 \times 10^{-4}$ kg · m^2 and $B_m = 0$.

SP6.4-1. Calculate the efficiency for $\phi_v = 0$ given in Example 6.A. [81.7%]
SP6.4-2. Obtain (6.4-3) and (6.4-4).

6.5   Field-Oriented Induction Motor Drive

The goal of field orientation of an induction machine is fast torque response, which is accomplished with a controlled drive inverter [5, 6]. Unfortunately, the necessary control is involved since not only is a torque control involved but it is also necessary to control the stator variables to orient the maximum value of the rotor field, mmf_r^e, in the q axis, orthogonal with the d-axis component of the stator field, mmf_s^e. To do all this, we have two things we can control, the stator applied voltage and the slip frequency. Although the principle of field orientation is quite straightforward, the implementation of the control is not. In particular, the voltage equations in the synchronous rotating reference frame are used to determine the values of the substitute currents necessary to ensure field orientation and these values are then transformed from the synchronous reference frame to the stationary reference frame to determine the waveforms of the stator currents and applied voltages necessary to produce the synchronous reference frame variables that will provide field orientation. This is achieved by controlling the output voltage of the drive inverter to shape the waveforms of the actual stator currents to that determined necessary to provide field orientation. In this section, we will not become involved with the details of implementing this control; instead, we will assume that the control is functioning perfectly and focus on the performance of an induction machine with field orientation. In other words, in this section, our focus will be on what the control does and not how it does it. Also, we will consider the two-phase device since this reduces our work and the extension to three-phase is straightforward.

The voltage and flux linkage equations of a two-phase single-fed induction machine in the synchronously rotating reference frame from Chapter 3 are

$$v_{qs}^e = r_s i_{qs}^e + \omega_e \lambda_{ds}^e + p\lambda_{qs}^e \tag{6.5-1}$$

$$v_{ds}^e = r_s i_{ds}^e - \omega_e \lambda_{qs}^e + p\lambda_{ds}^e \tag{6.5-2}$$

$$0 = r_r' i_{qr}'^e + (\omega_e - \omega_r)\lambda_{dr}'^e + p\lambda_{qr}'^e \tag{6.5-3}$$

$$0 = r_r' i_{dr}'^e - (\omega_e - \omega_r)\lambda_{qr}'^e + p\lambda_{dr}'^e \tag{6.5-4}$$

where

$$\lambda_{qs}^e = L_{ss} i_{qs}^e + L_{ms} i_{qr}'^e \tag{6.5-5}$$

$$\lambda_{ds}^e = L_{ss} i_{ds}^e + L_{ms} i_{dr}'^e \tag{6.5-6}$$

$$\lambda_{qr}'^e = L_{rr}' i_{qr}'^e + L_{ms} i_{qs}^e \tag{6.5-7}$$

$$\lambda_{dr}'^e = L_{rr}' i_{dr}'^e + L_{ms} i_{ds}^e \tag{6.5-8}$$

The torque may be expressed as

$$T_e = \frac{P}{2}\left(\lambda'^e_{qr} i'^e_{dr} - \lambda'^e_{dr} i'^e_{qr}\right) \tag{6.5-9}$$

The aim is to select the applied stator voltages so that $\lambda'^e_{qr} = 0$, whereupon (6.5-9) becomes

$$T_e = -\frac{P}{2}\lambda'^e_{dr} i'^e_{qr} \tag{6.5-10}$$

Making λ'^e_{qr} zero does not mean that i'^e_{qr} will also be zero; instead, from (6.5-7), we see that with λ'^e_{qr} zero, then

$$i'^e_{qr} = -\frac{L_{ms}}{L'_{rr}} i^e_{qs} \tag{6.5-11}$$

Thus, if i'^e_{qr} is (6.5-11), then $\lambda'^e_{dr} = 0$ and T_e is (6.5-10).

From (6.5-10), we see that if λ'^e_{dr} is constant, then torque is proportional to i^e_{qs} as given by (6.5-11). Now, let us control i'^e_{dr} to zero. Why? Well, if i'^e_{dr} is zero, the rotor poles will be positioned completely in the q axis; however, controlling i'^e_{dr} to zero does not mean that λ'^e_{dr} is zero; from (6.5-8) with $i'^e_{dr} = 0$, then

$$\lambda'^e_{dr} = L_{ms} i^e_{ds} \tag{6.5-12}$$

which will be constant if i^e_{ds} is held constant. If (6.5-11) and (6.5-12) are substituted into (6.5-10), the torque may be expressed as

$$T_e = \frac{P}{2}\frac{L^2_{ms}}{L'_{rr}} i^e_{qs} i^e_{ds} \tag{6.5-13}$$

Note that we have T_e in terms of stator-related currents and if i^e_{ds} is held constant, T_e is directly proportional to i^e_{qs}. This equation is used to control i^e_{qs} when T_e is commanded and with i^e_{ds} held constant, generally at its rated value.

We have eliminated (6.5-4) and since λ'^e_{qr} is zero, then so must be $p\lambda'^e_{qr}$, and if (6.5-11) is substituted for i'^e_{qr} and (6.5-12) for λ'^e_{dr} into (6.5-3), the angular velocity of the slip may be calculated as

$$(\omega_e - \omega_r)_{calc} = \frac{r'_r}{L'_{rr}}\frac{i^e_{qs}}{i^e_{ds}} \tag{6.5-14}$$

which we will also denote as $(\omega_s)_{calc}$ for compactness. This equation is used to control slip frequency. Note that (6.5-14), like (6.5-13), is in terms of stator-related currents. We have set forth the basic relationships for field orientation control which in effect positions the rotor poles along the q axis (6.5-11), orthogonal with the d axis (I_{ds}^e) or at $\theta_{esi}(0)$ with I_{qs}^e and I_{ds}^e or $\tilde{I}_{as}$.

Our purpose now is to consider the steady-state performance of an induction motor assuming that the field orientation is functioning properly. Although this idealized approach is an oversimplification of the control challenges involved, it helps to give insight to the basic features of field orientation; however, since the stator currents are commanded (controlled), the electric transients are minimized. Therefore, we will find that steady-state, "ideally" controlled operation and the actual field orientation drive with harmonics neglected are very similar.

Let us take a minute to express the steady-state torque assuming field orientation is functioning properly. We can express I_{qs}^e and I_{ds}^e, respectively, as

$$I_{qs}^e = \sqrt{2}I_s \cos\theta_{esi}(0) \tag{6.5-15}$$

$$I_{ds}^e = -\sqrt{2}I_s \sin\theta_{esi}(0) \tag{6.5-16}$$

where in (6.5-15) and (6.5-16), $\omega = \omega_e$ and $\theta(0) = 0$. Substituting (6.5-15) and (6.5-16) into (6.5-13) yields the expression for torque with field orientation as

$$
\begin{aligned}
T_e &= -\frac{P}{2}\frac{L_{ms}^2}{L_{rr}'}\left[\sqrt{2}I_s \cos\theta_{esi}(0)\right]\left[\sqrt{2}I_s \sin\theta_{esi}(0)\right] \\
&= -\frac{P}{2}\frac{L_{ms}^2}{L_{rr}'}I_s^2 \sin 2\theta_{esi}(0)
\end{aligned}
\tag{6.5-17}
$$

With field orientation and harmonics neglected, the flux-linkages in the synchronous reference frame are zero or constants, whereupon, ideally the time rate of change is zero. Therefore, with ideal functioning field orientation $p\tilde{i}_{as}$ is zero and the instantaneous phasor equations become the steady-state phasor equations. The only dynamic feature is the relationship between rotor speed and torque given by (3.6-14).

We will work with the steady-state equations and modify V_{qs}^e and V_{ds}^e to account for the results of field orientation action on $I_{qr}'^e$ and $I_{dr}'^e$ variables given earlier, thereby, making it necessary to work only with the stator phasor voltage equations. Substituting $I_{qr}'^e$, in terms of the I_{qs}^e from (6.5-11), into (6.5-5) yields

$$\lambda_{qs}^e = \left(\frac{L_{ss}}{L_{ms}} - \frac{L_{ms}}{L_{rr}'}\right)L_{ms}I_{qs}^e \tag{6.5-18}$$

We can now express steady-state V_{qs}^e and V_{ds}^e. From (6.5-1) and (6.5-2) with $p\lambda = 0$, λ_{ds}^e from (6.5-6), with $I_{dr}'^e$ zero, and (6.5-18) for λ_{qs}^e, we have

$$V_{qs}^e = r_s I_{qs}^e + \omega_e L_{ss} I_{ds}^e \tag{6.5-19}$$

$$V_{ds}^e = r_s I_{ds}^e - \omega_e \left(\frac{L_{ss}}{L_{ms}} - \frac{L_{ms}}{L_{rr}'} \right) L_{ms} I_{qs}^e \qquad (6.5\text{-}20)$$

Let us take a minute to express $\tilde{V}_{as}$ for steady-state operation. Substituting (6.5-19) and (6.5-20) into (3.8-9) yields $\sqrt{2}\,\tilde{V}_{as}$ with field orientation, that is,

$$
\begin{aligned}
\sqrt{2}\tilde{V}_{as} &= V_{qs}^e - jV_{ds}^e \\
&= r_s I_{qs}^e + \omega_e L_{ss} I_{ds}^e - j\left(r_s I_{ds}^e - \omega_e K_L L_{ms} I_{qs}^e \right) \\
&= r_s \left(I_{qs}^e - j I_{ds}^e \right) + \omega_e L_{ss} I_{ds}^e + j\omega_e K_L L_{ms} I_{qs}^e
\end{aligned}
\qquad (6.5\text{-}21)
$$

where

$$K_L = \frac{L_{ss}}{L_{ms}} - \frac{L_{ms}}{L_{rr}'} \qquad (6.5\text{-}22)$$

If now we add and subtract $\omega_e K_L L_{ms} I_{ds}^e$ on the right-hand side of (6.5-21), we can express $\tilde{V}_{as}$ as

$$\tilde{V}_{as} = (r_s + j\omega_e K_L L_{ms})\tilde{I}_{as} + \frac{1}{\sqrt{2}}\omega_e(L_{ss} - K_L L_{ms})I_{ds}^e \qquad (6.5\text{-}23)$$

Substituting (6.5-16) for I_{ds}^e into (6.5-23) yields

$$\tilde{V}_{as} = (r_s + j\omega_e K_L L_{ms})\tilde{I}_{as} - \omega_e(L_{ss} - K_L L_{ms})I_s \sin\theta_{esi}(0) \qquad (6.5\text{-}24)$$

Solving (6.5-13) for I_{qs}^e yields

$$I_{qs}^e = \frac{2}{P}\frac{L_{rr}'}{L_{ms}^2}\frac{T_e}{I_{ds}^e} \qquad (6.5\text{-}25)$$

We are also going to use (6.5-14), but let us wait just a minute before we get to that.

Although we are not going to become involved in the details of the field orientation, the block diagram shown in Fig. 6.5-1 depicts the basic principles of the control. Therein, the input variables containing an asterisk are commanded variables. That is, if the machine is to operate with rated load torque and rated frequency, then T_e^* would be rated torque and I_{ds}^{e*} would be constant and generally selected to be the value of I_{ds}^e for rated conditions and maintained at that value regardless of T_e^*. Now, (6.5-14), (6.5-25), and (6.5-26) which we will get to in a moment, impose field orientation control on the currents, i_{as}^* and i_{bs}^*. The inverter forms the voltages v_{as} and v_{bs} to ensure i_{as} and i_{bs} are i_{as}^* and i_{bs}^*, respectively.

We have I_{qs}^{e*} and I_{ds}^{e*} and, from (6.5-14), we can calculate $\omega_e - \omega_r$, which we are calling $(\omega_s)_{calc}$ in Fig. 6.5-1 (the slip angular velocity) and/or $(\omega_e - \omega_r)_{calc}$ since it is

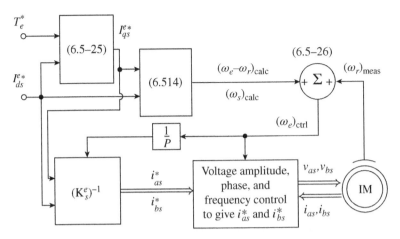

Figure 6.5-1 Block diagram depicting field-oriented control principles. Note:
$I_{qs}^e = \sqrt{2}\, I_s \cos\theta_{esi}(0)$ and $I_{ds}^e = -\sqrt{2}\, I_s \sin\theta_{esi}(0)$.

calculated from commanded inputs I_{qs}^{e*} and I_{ds}^{e*} and machine parameters. Now, $(\omega_s)_{calc}$ is the slip necessary for field orientation with the commanded stator-related currents. Now, $(\omega_r)_{meas}$ is the measured electrical angular velocity of the rotor and it is added to $(\omega_s)_{calc}$. Thus, from Fig. 6.5-1,

$$
\begin{aligned}
(\omega_e)_{ctrl} &= (\omega_s)_{calc} + (\omega_r)_{meas} \\
&= (\omega_e - \omega_r)_{calc} + (\omega_r)_{meas}
\end{aligned}
\tag{6.5-26}
$$

where $(\omega_e)_{ctrl}$ is the electrical angular velocity of the controlled stator voltages applied to the machine (v_{as} and v_{bs} of Fig. 6.5-1) for the rotor magnetic field to be oriented in the q axis. There is no direct control of speed; however, if we are operating in the steady state and the torque load increases, $(\omega_r)_{meas}$ would decrease which would decrease ω_e until $(\omega_e - \omega_r)_{calc}$ was established. Again, $(\omega_e - \omega_r)_{calc}$ is the slip angular velocity, calculated from the commanded T_e^* and I_{ds}^{e*} by (6.5-14). Assuming the parameters remain constant with temperature change, then $(\omega_e - \omega_r)_{calc}$ will change only if T_e^* and/or I_{ds}^{e*} are changed.

Example 6.E   Torque Control

The parameters of the two-phase two-pole 5-hp 110-V 60-Hz induction machine are

$$
r_s = 0.295\,\Omega \qquad L_{ms} = 35.15\,\text{mH} \qquad r_r' = 0.201\,\Omega
$$
$$
L_{ls} = 0.944\,\text{mH} \qquad\qquad\qquad L_{lr}' = 0.944\,\text{mH}
$$

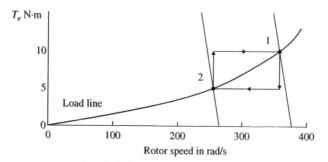

Figure 6.E-1 Operation of induction motor drive with field orientation for step change in T_e^*.

The machine is equipped with field orientation. The rated torque is 10 N · m and the rated power factor angle is −30°. This is Operating Point 1 of Fig. 6.E-1 which shows the torque versus speed trajectory for switching the command from 10 N · m to 5 N · m and then back to 10 N · m. From (6.5-17),

$$T_e = -\frac{P}{2}\frac{L_{ms}^2}{L'_{rr}}I_s^2 \sin 2\theta_{esi}(0)$$

$$10 = -\left(\frac{2}{2}\right)\frac{(35.15)^2 \times 10^{-3}}{(35.15 + 0.944)}I_s^2 \sin 2(-30°) \tag{6E-1}$$

Solving for I_s yields

$$I_s = \left[\frac{(10)(35.15 + 0.944)}{(35.15)^2 \times 10^{-3} \sin 60°}\right]^{1/2}$$
$$= 18.37 \text{ A} \tag{6E-2}$$

Therefore,

$$\tilde{I}_{as} = 18.37\underline{/-30°} \text{ A} \tag{6E-3}$$

Now,

$$L_{ss} = L_{ls} + L_{ms}$$
$$= (35.15 + 0.944) \times 10^{-3} \tag{6E-4}$$
$$= 36.094 \text{ mH}$$

From (6.5-22),

$$K_L = \frac{L_{ss}}{L_{ms}} - \frac{L_{ms}}{L'_{rr}} = \frac{36.094}{35.15} - \frac{35.15}{36.094} = 0.0531 \tag{6E-5}$$

From (6.5-23),

$$\tilde{V}_{as} = (r_s + j\omega_e K_L L_{ms})\tilde{I}_{as} - \omega_e(L_{ss} - K_L L_{ms})I_s \sin\theta_{esi}(0)$$
$$= [0.295 + j377(0.0531)(35.15 \times 10^{-3})]18.37\underline{/-30°}$$
$$- 377(36.094 - 0.0531 \times 35.15) \times 10^{-3}(18.37)(-0.5)$$
$$= 129.7 + j8.5 = 130\underline{/3.74°} \text{ V}$$

(6E-6)

This voltage exceeds rated (110 V); however, the current is rated. We will continue. The rotor speed is calculated from (6.5-14):

$$\omega_e - \omega_r = \frac{r'_r}{L'_{rr}} \frac{I^e_{qs}}{I^e_{ds}}$$

(6E-7)

where

$$I^e_{qs} = \sqrt{2}I_s \cos\theta_{esi}(0)$$
$$= \sqrt{2}\,18.37\cos(-30°) = 22.5 \text{ A}$$

(6E-8)

$$I^e_{ds} = -\sqrt{2}\,18.37\sin\theta_{esi}(0)$$
$$= \sqrt{2}\,18.37\cos(-30°) = 22.5 \text{ A}$$
$$- \sqrt{2}\,18.37\sin(-30°) = 12.99 \text{ A}$$

(6E-9)

From (6.5-14),

$$\omega_e - \omega_r = \frac{r'_r}{L'_{rr}} \frac{I^e_{qs}}{I^e_{ds}}$$
$$= \frac{0.201}{36.09 \times 10^{-3}} \frac{22.5}{12.99} = 9.6 \text{ r/s}$$
$$-\omega_r = -\omega_e + 9.6$$
$$\omega_r = 377 - 9.6 = 367.4 \text{ r/s}$$

(6E-10)

The drive is originally operating at Operating Point 1 of Fig. 6.E-1. The load line is

$$T_L = K\omega_r^2$$

(6E-11)

where for $T_e = 10$ N · m at Operating Point 1 with $\omega_r = 367.4$ rad/s.

$$K = \frac{10}{(367.4)^2} = 7.41 \times 10^{-5} \text{ N · m · s}^2$$

(6E-12)

The phasor diagram for Operating Point 1 is given in Fig. 6.E-2.

We see that $\tilde{I}'_{ar}$ is oriented orthogonal to I^e_{ds} and the torque is 10 N · m and this has been done by controlling $\tilde{V}_{as}$ and $\omega_e - \omega_r$. The phasor diagram shown in Fig. 6.E-2 should be compared to Fig. 3.D-2 which is for uncontrolled rotor field orientation.

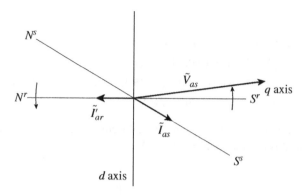

Figure 6.E-2 Phasor diagram for Operating Point 1.

The commanded torque is changed from 10 N · m to 5 N · m. The motor slows and after the mechanical dynamics subside, the new Operating Point is at 2 on Fig. 6.E-1. At this point the new speed from (6D-4) is

$$T_L = K\omega_r^2$$
$$5 = 7.41 \times 10^{-5}\omega_r^2$$
$$\omega_r = \left(\frac{5}{7.41 \times 10^{-5}}\right)^{1/2} = 259.8 \text{ rad/s} \tag{6E-13}$$

We could liken this scenario to reducing the accelerator of an electric car by one half. Now let us return to Operating Point 1, the accelerator is depressed and we return to Operating Point 1 as shown in Fig. 6.E-1.

Let us calculate the applied voltage of Operating Point 2.

$$\tilde{V}_{as} = (r_s + j\omega_e K_L L_{ms})\tilde{I}_{as} - \omega_e(L_{ss} - K_L L_{ms})I_s \sin\theta_{esi}(0) \tag{6E-14}$$

Now for $T_e^* = 5 \text{ N} \cdot \text{m}$ $\omega_e - \omega_r = 4.8 \text{ rad/s}$, therefore

$$\omega_e = 259.8 + 4.8 = 264.6 \text{ rad/s} \tag{6E-15}$$

Also, $\tilde{I}_{as} = 12.2\,\underline{/-49.1°}$ and for a two-phase motor $r_s = 0.295\,\Omega$, $L_{ss} = 36.094 \text{ mH}$, $L_{ms} = 35.15 \text{ mH}$, and $K_L = 0.053$.

Substituting into (6E-14),

$$\tilde{V}_{as} = \left[0.295 + j264.6(0.053)(35.151 \times 10^{-3})\right] 12.2\,\underline{/-49.1°}$$
$$- 264.6(36.094 - 0.053 \times 35.15) \times 10^{-3}(12.2)(\sin -49.1°) \tag{6E-16}$$
$$= 0.5744\,\underline{/59.1°}\ 12.2\,\underline{/-49.1°} + 83.51\,\underline{/0°}$$
$$= 90 + j1.15 = 90\,\underline{/0.732°}\text{ V}$$

SP6.5-1. Calculate I_{as} for $T_L = 5\,\text{N} \cdot \text{m}$ at Operating Point 2. $[\tilde{I}_{as} = 12.1\,\underline{/-49.2}\,\text{A}]$

SP6.5-2. Draw the phasor diagram for Operating Point 2 of Fig. 6.E-1.

SP6.5-3. What changes are necessary for a three-phase induction motor. $[L_{ms}$ is replaced with L_{Ms} and T_e are multiplied by $\frac{3}{2}]$

6.6 Problems

1 It is found that $\lambda''_m = 0.1\,\text{V} \cdot \text{s/rad}$ for a permanent-magnet six-pole two-phase ac machine. Calculate the amplitude (peak value) of the open-circuit phase voltage measured when the rotor is turned at 60 revolutions per second (r/s).

2 Verify Fig. 6.2-1c.

3 In the analysis of the brushless dc motor, we have selected $\theta_r(0) = 0$, where $\sqrt{2}\tilde{F}_{as} = F^r_{qs} - jF^r_{ds}$. Express the relationship between these same variables if we had selected $\theta_r(0) = \dfrac{1}{2}\pi$.

4 Repeat Example 6.D for $\omega_{rm} = 600\,\text{r/min}$. Construct the phasor diagram showing $\tilde{V}_{as}, \tilde{E}_a, r_s\tilde{I}_{as}, j\omega_r L_{ss}\tilde{I}_{as}$, and $\tilde{I}_{as}$. Show the pole locations.

5 A four-pole two-phase permanent-magnet ac machine is driven by a mechanical source at $\omega_r = 3600\,\text{r/min}$. The open-circuit voltage across one of the phases is $50\,\text{V}$ (rms). (a) Calculate λ''_m. The mechanical source is removed and the following voltages are applied: $V_{as} = \sqrt{2}\,25\cos\theta_r$, $V_{bs} = \sqrt{2}\,25\sin\theta_r$, where $\theta_r = \omega_r t$. (b) Neglect friction ($B_m = 0$) and calculate the no-load rotor speed ω_r in rad/s.

6 The parameters of a two-pole three-phase permanent-magnet ac machine are $r_s = 4\,\Omega$, $\lambda''_m = 0.07\,\text{V} \cdot \text{s/rad}$, $L_{ls} = 1\,\text{mH}$, and $L_{Ms} = 13.5\,\text{mH}$. Let $\theta_r = 200t$. Determine (a) $\tilde{I}_{as}$, (b) T_e, and (c) show the phasor diagram and locate the poles.

7 Repeat Prob. 6 with $\phi_{vMT/V}$.

8 Repeat Prob. 6 with $\phi_{vMT/A}$.

9 Plot v_{as}, v_{bs}, v_{cs}, v_{ab}, and v_{ng} for the 180° continuous-current inverter for $0 < \theta_r < 2\pi$ with the transistor switching shown in Fig. 4.4-1. Assume the inverter voltage v_i is constant.

10 Verify the rotor speed for Operating Point 1 of Example 6.E.

References

1 T. W. Nehl, F. A. Fouad, and N. A. Demerdash, "Digital Simulation of Power Conditioner-Machine Interaction for Electronically Commutated DC Permanent Magnet Machines," *IEEE Trans. On Magnetics*, Vol. **17**, November 1981, pp. 3284–3286.

2 T. M. Jahns, "Torque Production in Permanent-Magnet Synchronous Motor Drives with Rectangular Current Excitations," *IAS Conf. Rec.*, Vol. **IA-20**, October 1983, pp. 476–487.

3 P. C. Krause, O. Wasynczuk, T. C. O'Connell, and M. Hasan, *Introduction to Power and Drive Systems*, Wiley, Hoboken, NJ, USA, September 2017.

4 P. C. Krause, *Analysis of Electric Machinery*, McGraw-Hill Book Company, New York, NY, 1986.

5 A. M. Traynadlowski, *The Field Orientation Principle in Control of Induction Motors*, TU Braunschweig, Germany, 1994.

6 F. Blaschke, *Das Verfahren der Feldorientierung zur Regelung der Drehfeld-maschine*, Ph.D. Thesis, TU Braunschweig (Technische Universität Braunschweig), 1974.

7

Single-Phase Induction Motors

7.1 Introduction

Although the voltage and torque equations set forth in Chapter 3 for the induction machine are valid regardless of the mode of operation, we focused on balanced conditions. Single-phase induction motors are used in household washers, dryers, air conditioners, garbage disposals, etc., where only a single-phase source is available. Symmetrical two-phase induction motors are often used in these single-phase applications; however, in order to develop a starting torque, it is necessary to make the symmetrical two-phase induction motor think it is being supplied from a two-phase source or, at least, something that resembles a two-phase source. We will find that this can be accomplished by placing a capacitor in series with one of the stator phase windings until the rotor has accelerated to between 60 and 80% of rated operating speed, whereupon the series combination of the capacitor and the phase winding are disconnected from the source. The motor then operates with only one of its phases connected to the single-phase source. Hence, there are two common modes of unbalanced operation of a symmetrical two-phase induction motor when used as a single-phase device; first, during starting, the phase voltages applied are not a balanced two-phase set and the input impedance of one phase is different from the other owing to the series capacitor, and second, during near rated operation one stator phase is often open-circuited.

To analyze steady-state unbalanced operation of an induction machine, it is convenient to use the method of symmetrical components [1]. This method is introduced in the following section and used to analyze unbalanced stator voltages, unequal stator impedances, and an open-circuited stator phase since all of these modes of operation occur during operation of a single-phase induction motor.

It is interesting that the single-phase operation of an induction machine is somewhat the opposite of the synchronous machine where the stator has symmetrical

Introduction to Modern Analysis of Electric Machines and Drives, First Edition.
Paul C. Krause and Thomas C. Krause.
© 2023 The Institute of Electrical and Electronics Engineers, Inc.
Published 2023 by John Wiley & Sons, Inc.

excited windings and the rotor has unsymmetrical windings. In the case of the single-phase operation of a symmetrical induction machine, the rotor has "symmetrical" windings but the stator is unsymmetrically excited. Reference frame theory can be used to analyze the single-phase machine without using symmetrical components; however, this approach is very similar to using symmetrical components. This is more or less expected since the symmetrical component transformation comes from reference frame theory which comes from Tesla's rotating magnetic field [2].

In this chapter, steady-state and dynamic characteristics are illustrated for single-phase applications; however, only the symmetrical two-phase induction motor is considered in this chapter. Actually the unsymmetrical two-phase induction motor or the so-called *split-phase* induction motor is often used rather than its symmetrical two-phase cousin [3]. Although the last section of this chapter is devoted to a brief discussion of the split-phase machine, it is not analyzed. We have chosen to illustrate the salient operating features of single-phase induction machines by using the symmetrical device since it is far easier to analyze than the split-phase machine.

7.2 Symmetrical Components

We must deal with unbalanced conditions in the analysis of steady-state operation of the symmetrical two-phase induction motor when used in single-phase applications. This can be accomplished by using what is referred to as the method of symmetrical components. C. L. Fortescue [1] published the method of symmetrical components for the purpose of analyzing unbalanced multiphase systems. Since that time, this method has been extended, modified, and used widely, sometimes inappropriately. Nevertheless, it is a powerful analytical tool for steady-state unbalanced operation of symmetrical systems; even though the derivation and the procedure for applying this method often seemed to be without theoretical basis. It has been shown, however, that reference frame theory provides a rigorous derivation of the method of symmetrical components and sets clear guidelines for its application [2]. In this section, we will describe the concept of symmetrical components without derivation and establish the equations necessary to conduct the analysis of unbalanced operation. A theoretical justification of the concept of symmetrical components using reference frame theory, which is based on Tesla's rotating magnet field, is found in [2, 3].

For convenience, Fig. 3.2-3 of the two-pole two-phase symmetrical induction machine is repeated in Fig. 7.2-1. The method of symmetrical components allows us to represent an unbalanced two-phase set as two balanced sets or an unbalanced three-phase set as two balanced sets and a single phase. The balanced sets are referred to as the *positive-sequence* and *negative-sequence* components and the

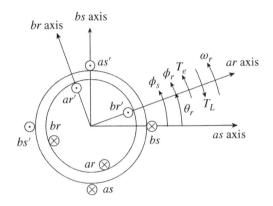

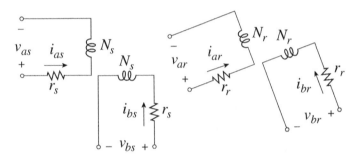

Figure 7.2-1 A two-pole, two-phase symmetrical induction machine.

single phase is called the *zero-sequence* component. Here, we will deal only with a two-phase system. Also, the method of symmetrical components is valid only for steady-state conditions.

In general, an unbalanced two-phase set may be expressed as

$$F_{as} = \sqrt{2}\, F_a \cos\left[\omega_e t + \theta_{efa}(0)\right] \tag{7.2-1}$$

$$F_{bs} = \sqrt{2}\, F_b \sin\left[\omega_e t + \theta_{efb}(0)\right] \tag{7.2-2}$$

We realize that (7.2-1) and (7.2-2) form a balanced set if $F_a = F_b$ and $\theta_{efa}(0) = \theta_{efb}(0)$. It is convenient for us to work with qs^s and ds^s variables rather than as and bs variables. Recall that with $\omega = \omega_e$ and $\theta(0) = 0$, $f_{as} = f_{qs}^s$ and $f_{bs} = -f_{ds}^s$ or $\tilde{F}_{as} = \tilde{F}_{qs}^s$ and $\tilde{F}_{bs} = -\tilde{F}_{ds}^s$ in the steady state.

It can be shown that an unbalanced two-phase set may be broken up into two balanced sets as

$$\tilde{F}_{qs}^s = \tilde{F}_{qs+}^s + \tilde{F}_{qs-}^s \tag{7.2-3}$$

$$\tilde{F}_{ds}^s = \tilde{F}_{ds+}^s + \tilde{F}_{ds-}^s \qquad (7.2\text{-}4)$$

Here, we are departing somewhat from tradition. Rather than using qs^s and ds^s variables, it has been customary to use as and bs variables and, hence, (7.2-3) and (7.2-4) are written, respectively, as $\tilde{F}_{as} = \tilde{F}_{as+} + \tilde{F}_{as-}$ and $\tilde{F}_{bs} = \tilde{F}_{bs+} + \tilde{F}_{bs-}$. Later, we will substitute $\tilde{F}_{as}$ for $\tilde{F}_{qs}^s$ and $\tilde{F}_{bs}$ for $-\tilde{F}_{ds}^s$; however, for purposes of convenience, we will continue the break from tradition and use $\tilde{F}_{qs+}^s$, $\tilde{F}_{qs-}^s$, $\tilde{F}_{ds+}^s$, and $\tilde{F}_{ds-}^s$ rather than $\tilde{F}_{as+}$, $\tilde{F}_{as-}$, $-\tilde{F}_{bs+}$ and $-\tilde{F}_{bs-}$, respectively.

In (7.2-3) and (7.2-4), $\tilde{F}_{qs+}^s$ and $\tilde{F}_{ds+}^s$ form the balanced, positive-sequence set, where

$$\tilde{F}_{ds+}^s = j\tilde{F}_{qs+}^s \qquad (7.2\text{-}5)$$

The negative-sequence set is $\tilde{F}_{qs-}^s$ and $\tilde{F}_{ds-}^s$, where

$$\tilde{F}_{ds-}^s = -j\tilde{F}_{qs-}^s \qquad (7.2\text{-}6)$$

Both the positive- and negative-sequence sets are balanced, but in the case of the positive-sequence set $\tilde{F}_{ds+}^s$ leads $\tilde{F}_{qs+}^s$ ($\tilde{F}_{bs+}$ lags $\tilde{F}_{as+}$) by 90°, whereas in the case of the negative-sequence set $\tilde{F}_{ds-}^s$ lags $\tilde{F}_{qs-}^s$ ($\tilde{F}_{bs-}$ leads $\tilde{F}_{as-}$). Why is one set called the positive sequence and the other the negative sequence? Well, probably the best explanation is to point out that positive-sequence currents flowing in the stator windings (Fig. 7.2-1) will produce an air-gap mmf which rotates counterclockwise whereas negative-sequence currents will produce an air-gap mmf which rotates clockwise. (Do not forget that $\tilde{F}_{bs+} = -\tilde{F}_{ds+}$ and $\tilde{F}_{bs-} = -\tilde{F}_{ds-}$.) It should be apparent that, if $\tilde{F}_{as}$ and $\tilde{F}_{bs}$ are balanced with $\tilde{F}_{bs} = -j\tilde{F}_{as}$, then the negative-sequence variables ($\tilde{F}_{qs-}^s$ and $\tilde{F}_{ds-}^s$) would not exist.

Substituting (7.2-5) and (7.2-6) into (7.2-3) and (7.2-4) yields

$$\begin{bmatrix} \tilde{F}_{qs}^s \\ \tilde{F}_{ds}^s \end{bmatrix} = \begin{bmatrix} 1 & 1 \\ j & -j \end{bmatrix} \begin{bmatrix} \tilde{F}_{qs+}^s \\ \tilde{F}_{qs-}^s \end{bmatrix} \qquad (7.2\text{-}7)$$

Now, let us substitute $\tilde{F}_{as}$ for $\tilde{F}_{qs}^s$ and $\tilde{F}_{bs}$ for $-\tilde{F}_{ds}^s$, whereupon

$$\begin{bmatrix} \tilde{F}_{as} \\ \tilde{F}_{bs} \end{bmatrix} = \begin{bmatrix} 1 & 1 \\ -j & j \end{bmatrix} \begin{bmatrix} \tilde{F}_{qs+}^s \\ \tilde{F}_{qs-}^s \end{bmatrix} \qquad (7.2\text{-}8)$$

Solving for $\tilde{F}^s_{qs+}$ and $\tilde{F}^s_{qs-}$ yields

$$\begin{bmatrix} \tilde{F}^s_{qs+} \\ \tilde{F}^s_{qs-} \end{bmatrix} = \frac{1}{2}\begin{bmatrix} 1 & j \\ 1 & -j \end{bmatrix}\begin{bmatrix} \tilde{F}_{as} \\ \tilde{F}_{bs} \end{bmatrix} \tag{7.2-9}$$

The symmetrical-component transformation matrix is defined from (7.2-9) as

$$\mathbf{S} = \frac{1}{2}\begin{bmatrix} 1 & j \\ 1 & -j \end{bmatrix} \tag{7.2-10}$$

and

$$(\mathbf{S})^{-1} = \begin{bmatrix} 1 & 1 \\ -j & j \end{bmatrix} \tag{7.2-11}$$

Example 7.A Symmetrical-Component Transformation

The steady-state variables of an unbalanced two-phase system are

$$\tilde{F}_{as} = 1\underline{/45°} \tag{7A-1}$$

$$\tilde{F}_{bs} = \frac{1}{2}\underline{/-120°} \tag{7A-2}$$

Calculate $\tilde{F}^s_{qs+}$ and $\tilde{F}^s_{qs-}$. Substituting into (7.2-9),

$$\begin{bmatrix} \tilde{F}^s_{qs+} \\ \tilde{F}^s_{qs-} \end{bmatrix} = \frac{1}{2}\begin{bmatrix} 1 & j \\ 1 & -j \end{bmatrix}\begin{bmatrix} 1\underline{/45°} \\ \frac{1}{2}\underline{/-120°} \end{bmatrix} \tag{7A-3}$$

From which

$$\begin{aligned} \tilde{F}^s_{qs+} &= \frac{1}{2}\left(1\underline{/45°} + j\frac{1}{2}\underline{/-120°}\right) \\ &= \frac{1}{2}\underline{/45°} + \frac{1}{4}\underline{/-30°} \\ &= 0.570 + j0.229 = 0.614\underline{/21.9°} \end{aligned} \tag{7A-4}$$

$$\begin{aligned} \tilde{F}^s_{qs-} &= \frac{1}{2}\left(1\underline{/45°} - j\frac{1}{2}\underline{/-120°}\right) \\ &= \frac{1}{2}\underline{/45°} - \frac{1}{4}\underline{/-30°} \\ &= 0.137 + j0.479 = 0.498\underline{/74.0°} \end{aligned} \tag{7A-5}$$

SP7.2-1. $\tilde{F}_{as} = \tilde{F}_{bs} = F_s\underline{/0°}$, determine $\tilde{F}^s_{qs+}$ and $\tilde{F}^s_{qs-}$. $\left[\tilde{F}^s_{qs-} = \tilde{F}^{s*}_{qs+} = \left(\sqrt{2}/2\right)F_s\underline{/45°}\right.$, where the asterisk denotes the conjugate]

SP7.2-2. Express $\tilde{F}_{bs}$ in terms of $\tilde{F}_{as}$ so that the positive-sequence component is zero. $\left[\tilde{F}_{bs} = j\tilde{F}_{as}\right]$

SP7.2-3. $\tilde{F}_{ds+}^{s} = \tilde{F}_{ds-}^{s}$ determine $\tilde{F}_{as}$ and $\tilde{F}_{bs}$. $\left[\tilde{F}_{as} = 0, \tilde{F}_{bs} = -2j\tilde{F}_{qs+}^{s}\right]$

7.3 Analysis of Unbalanced Modes of Operation

The rotor windings of the two-phase induction machine are short-circuited (squirrel-cage windings). In this analysis, the rotor speed is assumed constant, whereupon, the equations which describe constant-speed operation are linear and the principle of superposition applies. Therefore, the substitute variables ($F_{qr}^{\prime s}$ and $F_{dr}^{\prime s}$) for the rotor variables may be broken up into positive- and negative-sequence quantities in the same manner as F_{qs}^{s} and F_{ds}^{s}. In particular, we can write

$$\tilde{F}_{qr}^{\prime s} = \tilde{F}_{qr+}^{\prime s} + \tilde{F}_{qr-}^{\prime s} \tag{7.3-1}$$

$$\tilde{F}_{dr}^{\prime s} = \tilde{F}_{dr+}^{\prime s} + \tilde{F}_{dr-}^{\prime s} \tag{7.3-2}$$

and

$$\tilde{F}_{dr+}^{\prime s} = j\tilde{F}_{qr+}^{\prime s} \tag{7.3-3}$$

$$\tilde{F}_{dr-}^{\prime s} = -j\tilde{F}_{qr-}^{\prime s} \tag{7.3-4}$$

Armed with this information, let us see what we can derive in the way of voltage equations. From Chapter 3, the voltage equations for steady-state operation of a two-phase induction machine in the stationary reference frame.

$$\begin{bmatrix} \tilde{V}_{qs}^{s} \\ \tilde{V}_{ds}^{s} \\ \tilde{V}_{qr}^{\prime s} \\ \tilde{V}_{dr}^{\prime s} \end{bmatrix} = \begin{bmatrix} r_s + j\omega_e L_{ss} & 0 & j\omega_e L_{ms} & 0 \\ 0 & r_s + j\omega_e L_{ss} & 0 & j\omega_e L_{ms} \\ j\omega_e L_{ms} & -\omega_r L_{ms} & r_r^{\prime} + j\omega_e L_{rr}^{\prime} & -\omega_r L_{rr}^{\prime} \\ \omega_r L_{ms} & j\omega_e L_{ms} & \omega_r L_{rr}^{\prime} & r_r^{\prime} + j\omega_e L_{rr}^{\prime} \end{bmatrix} \begin{bmatrix} \tilde{I}_{qs}^{s} \\ \tilde{I}_{ds}^{s} \\ \tilde{I}_{qr}^{\prime s} \\ \tilde{I}_{dr}^{\prime s} \end{bmatrix} \tag{7.3-5}$$

Since for constant rotor speed ω_r, the voltage equations are linear and superposition applies. Thus, we can express the four equations of (7.3-5) twice; once for the positive-sequence variables and once for the negative-sequence variables. This gives two sets of four equations each. One set relates the positive-sequence voltages and currents, the other relates the negative-sequence voltages and currents.

However, since $\tilde{F}^s_{ds+} = j\tilde{F}^s_{qs+}$, $\tilde{F}'^s_{dr+} = j\tilde{F}'^s_{qr+}$, $\tilde{F}^s_{ds-} = -j\tilde{F}^s_{qs-}$, and $\tilde{F}'^s_{dr-} = -j\tilde{F}'^s_{qr-}$, the eight equations can be reduced back to four. If the d-variables are expressed in terms of the q-variables, the four equations are

$$
\begin{bmatrix}
\tilde{V}^s_{qs+} \\
\tilde{V}'^s_{qr+} \\ \dfrac{\ }{s} \\
\tilde{V}^s_{qs-} \\
\tilde{V}'^s_{qr-} \\ \dfrac{\ }{2-s}
\end{bmatrix}
=
\begin{bmatrix}
r_s + jX_{ss} & jX_{ms} & 0 & 0 \\
jX_{ms} & \dfrac{r'_r}{s} + jX'_{rr} & 0 & 0 \\
0 & 0 & r_s + jX_{ss} & jX_{ms} \\
0 & 0 & jX_{ms} & \dfrac{r'_r}{2-s} + jX'_{rr}
\end{bmatrix}
\begin{bmatrix}
\tilde{I}^s_{qs+} \\
\tilde{I}'^s_{qr+} \\
\tilde{I}^s_{qs-} \\
\tilde{I}'^s_{qr-}
\end{bmatrix}
\tag{7.3-6}
$$

where

$$X_{ss} = \omega_e(L_{ls} + L_{ms}) \tag{7.3-7}$$

$$X'_{rr} = \omega_e(L'_{lr} + L_{ms}) \tag{7.3-8}$$

$$X_{ms} = \omega_e L_{ms} \tag{7.3-9}$$

$$s = \frac{\omega_e - \omega_r}{\omega_e} \tag{7.3-10}$$

We realize that at any time we can change to the notation generally used by replacing $\tilde{F}^s_{qs+}$ with $\tilde{F}_{as+}$, $\tilde{F}'^s_{qr+}$ with $\tilde{F}'_{ar+}$, etc.

When we look at (7.3-6), we see that the positive- and negative-sequence variables are decoupled. From this, we might be led to believe that the positive- and negative-sequence variables may be considered separately regardless of the mode of operation of the induction motor. Although the voltage equations given by (7.3-6) provide a starting point, system constraints may cause the positive- and negative-sequence variables to be coupled. In the modes of operation which we will consider, we shall find that the sequence variables are decoupled when unbalanced voltages are applied to a symmetrical two-phase induction motor but coupled when an impedance is placed in series with one of the stator phase windings or when one of the stator phase windings is open-circuited.

The expression for the steady-state electromagnetic torque may be obtained by expressing it in terms of steady-state stationary reference frame currents.

$$T_e = \frac{P}{2} L_{ms} \left(I^s_{qs} I'^s_{dr} - I^s_{ds} I'^s_{qr} \right) \tag{7.3-11}$$

The instantaneous steady-state currents may each be expressed in terms of positive- and negative-sequence components. In particular, let

$$I^s_{qs} = \sqrt{2}\, I_{s+} \cos(\omega_e t + \phi_{s+}) + \sqrt{2}\, I_{s-} \cos(\omega_e t + \phi_{s-}) \tag{7.3-12}$$

$$I^s_{ds} = -\sqrt{2}\, I_{s+} \sin(\omega_e t + \phi_{s+}) + \sqrt{2}\, I_{s-} \sin(\omega_e t + \phi_{s-}) \tag{7.3-13}$$

$$I''^{s}_{qr} = \sqrt{2}\; I'_{r+}\; \cos\left(\omega_e t + \phi_{r+}\right) + \sqrt{2}\; I'_{r-}\; \cos\left(\omega_e t + \phi_{r-}\right) \tag{7.3-14}$$

$$I''^{s}_{dr} = -\sqrt{2}\; I'_{r+}\; \sin\left(\omega_e t + \phi_{r+}\right) + \sqrt{2}\; I'_{r-}\; \sin\left(\omega_e t + \phi_{r-}\right) \tag{7.3-15}$$

where the $+$ and $-$ subscripts denote positive- and negative-sequence quantities, respectively. If these expressions for the currents are substituted into (7.3-11) and with a few trigonometric identities, we can express the steady-state (constant-speed) torque as

$$\begin{aligned}
T_e = 2\left(\frac{P}{2}\right)L_{ms}[I_{s+}I'_{r+}\; \sin\left(\phi_{s+} - \phi_{r+}\right) - I_{s-}I'_{r-}\; \sin\left(\phi_{s-} - \phi_{r-}\right) \\
+ I_{s+}I'_{r-}\; \sin\left(2\omega_e t + \phi_{s+} + \phi_{r-}\right) - I_{s-}I'_{r+}\; \sin\left(2\omega_e t + \phi_{s-} + \phi_{r+}\right]
\end{aligned} \tag{7.3-16}$$

It is interesting that with the assumption of symmetrical rotor circuits, the electromagnetic torque during steady-state unbalanced operation is made up of a constant and a sinusoidal component which pulsates at twice the frequency of the stator variables. Recall that we have assumed that the steady-state stator variables contain only one frequency, ω_e. Multiple frequencies are treated in [2].

The above equation for torque may be expressed in terms of positive- and negative-sequence current phasors. After considerable work,

$$\begin{aligned}
T_e = 2\frac{P}{2}L_{ms}\Big\{ \text{Re}\left[j\left(\tilde{I}^{s*}_{qs+}\tilde{I}'^{s}_{qr+} - \tilde{I}^{s*}_{qs-}\tilde{I}'^{s}_{qr-}\right)\right] \\
+ \text{Re}\left[j\left(-\tilde{I}^{s}_{qs+}\tilde{I}'^{s}_{qr-} + \tilde{I}^{s}_{qs-}\tilde{I}'^{s}_{qr+}\right)\right]\cos 2\omega_e t \\
+ \text{Re}\left[\tilde{I}^{s}_{qs+}\tilde{I}'^{s}_{qr-} - \tilde{I}^{s}_{qs-}\tilde{I}'^{s}_{qr+}\right]\sin 2\omega_e t\Big\}
\end{aligned} \tag{7.3-17}$$

where the asterisk denotes the conjugate. The constant term [first term on right-hand side of (7.3-17)] is made up of the positive-sequence torque and the negative-sequence torque. The last two terms, which represent the pulsating torque component, could be combined; however, separate terms are somewhat more convenient. We note that the amplitude of the pulsating torque is related to the cross product of sequence currents.

7.3.1 Unbalanced Stator Voltages

During the starting period of a single-phase induction motor, a capacitor is placed in series with one of the windings. This type of unbalance, wherein the stator circuits appear unsymmetrical to the source because of the series capacitor, must be analyzed differently than for balanced conditions. We will consider the case of unbalanced source voltages applied to a symmetrical machine first and leave the series capacitor case for later.

Let us return to the voltage equations given by (7.3-6). We can apply these equations directly to solve for the sequence currents with unbalanced source voltages applied to the stator windings of a symmetrical machine. We need only to determine $\tilde{V}^s_{qs+}$ and $\tilde{V}^s_{qs-}$ from $\tilde{V}_{as}$ and $\tilde{V}_{bs}$ by (7.2-9). We know that, since the rotor windings are short-circuited, $\tilde{V}'^s_{qr+}$ and $\tilde{V}'^s_{qr-}$ are zero.

Since this unbalanced mode of operation is described by positive- and negative-sequence quantities which are decoupled, it is instructive to portray the four voltage equations given by (7.3-6) in equivalent-circuit form as shown in Fig. 7.3-1. The positive-sequence equivalent circuit is identical in form to that given for balanced conditions in Fig. 3.D-1 with the rotor windings short-circuited. This was expected. The negative-sequence equivalent circuit differs only in that the slip s is replaced by $2 - s$. Recall that the negative sequence voltages cause negative-sequence currents which cause a negatively rotating air-gap mmf. With respect to this negatively rotating air-gap mmf, the slip is $(\omega_e + \omega_r)/\omega_e$, which is $2 - (\omega_e - \omega_r)/\omega_e$ or $2 - s$. This line of reasoning is often used to obtain the negative-sequence equivalent circuit in place of a derivation.

Equation (7.3-6) or the equivalent circuits which come from (7.3-6) can be used to solve for the sequence currents. The steady-state electromagnetic torque can then be calculated by appropriate substitution of the sequence currents into (7.3-17). Since the positive- and negative-sequence circuits are decoupled, the positive- and negative-sequence torques may be expressed from (3.8-16). In particular, the positive-sequence torque, which is due to the product of the positive-sequence currents in the first term on the right-hand side of (7.3-17), may be expressed as

$$T_{e+} = \frac{2(P/2)\left(X^2_{ms}/\omega_e\right)r'_r s\left|\tilde{V}^s_{qs+}\right|^2}{\left[r_s r'_r + s\left(X^2_{ms} - X_{ss}X'_{rr}\right)\right]^2 + \left(r'_r X_{ss} + s r_s X'_{rr}\right)^2} \tag{7.3-18}$$

The negative-sequence torque, which is due to the product of the negative-sequence currents in the first term on the right-hand side of (7.3-17), may be expressed

$$T_{e-} = \frac{2(P/2)\left(X^2_{ms}/\omega_e\right)r'_r(2-s)\left|\tilde{V}^s_{qs-}\right|^2}{\left[r_s r'_r + (2-s)\left(X^2_{ms} - X_{ss}X'_{rr}\right)\right]^2 + \left[r'_r X_{ss} + (2-s)r_s X'_{rr}\right]^2} \tag{7.3-19}$$

Equation (7.3-18) was obtained from (3.8-16) with $\tilde{V}_{as}$ replaced by $\tilde{V}^s_{qs+}$ and (7.3-19) was obtained from (3.8-16) with $\tilde{V}_{as}$ replaced by $\tilde{V}^s_{qs-}$ and s replaced by $(2 - s)$. The average torque, $T_{e,\text{ave}}$, is the difference between the positive- and negative-sequence torques:

$$T_{e,\text{ave}} = T_{e+} - T_{e-} \tag{7.3-20}$$

Comparing the first two terms in (7.3-17) with (3.8-11), it is interesting to observe that, although torque, in general, is a nonlinear function of currents, we can use superposition to establish the total average torque by first calculating the positive- and negative-sequence currents from (7.3-6) or Fig. 7.3-1, then calculating individually the corresponding positive- and negative-sequence torques, and finally superimposing the results by using (7.3-20). However, it should be clear that, although superposition may be used to calculate the net average torque, the instantaneous torque (sum of average and pulsating components) cannot be calculated by using superposition since, from (7.3-17), the pulsating torque is related to the product of positive- and negative sequence currents.

Although we could express the amplitude of the pulsating torque in terms of the sequence voltages, the algebraic manipulations necessary to do so are a bit prohibitive. It is sufficient, for our purposes, to take a little closer look at the phasor relationship $\tilde{I}^s_{qs+} \tilde{I}'^s_{qr-} - \tilde{I}^s_{qs-} \tilde{I}'^s_{qr+}$ which is common to the second and third terms of (7.3-17). With the rotor winding short-circuited, we can express

$$\tilde{I}'^s_{qr+} = -\frac{jX_{ms}}{r'_r/s + jX'_{rr}} \tilde{I}^s_{qs+} \tag{7.3-21}$$

$$\tilde{I}'^s_{qr-} = -\frac{jX_{ms}}{r'_r/(2-s) + jX'_{rr}} \tilde{I}^s_{qs-} \tag{7.3-22}$$

These equations are obtained from the equivalent circuits given in Fig. 7.3-1. Note the similarity between (7.3-21) and (3.8-12). Utilizing (7.3-21) and (7.3-22), we can write

$$\tilde{I}^s_{qs+} \tilde{I}'^s_{qr-} - \tilde{I}^s_{qs-} \tilde{I}'^s_{qr+} = -jX_{ms} \tilde{I}^s_{qs+} \tilde{I}^s_{qs-} \frac{2(1-s)/[s(2-s)]}{[r'_r/(2-s) + jX'_{rr}](r'_r/s + jX'_{rr})} \tag{7.3-23}$$

If we express the sequence currents in terms of the sequence voltages, we can write (7.3-23) as

$$\tilde{I}^s_{qs+} \tilde{I}'^s_{qr-} - \tilde{I}^s_{qs-} \tilde{I}'^s_{qr+} = -jX_{ms} \frac{\tilde{V}^s_{qs+}}{Z_+} \frac{\tilde{V}^s_{qs-}}{Z_-} \frac{2(1-s)/[s(2-s)]}{[r'_r/(2-s) + jX'_{rr}](r'_r/s + jX'_{rr})} \tag{7.3-24}$$

where Z_+ and Z_- are the input impedances of the positive- and negative-sequence equivalent circuits (Fig. 7.3-1), respectively.

The form of (7.3-24) allows a somewhat more direct means of calculating the amplitude of the pulsating torque. It is interesting, however, to evaluate (7.3-24) for the condition where the rotor speed is zero. With $\omega_r = 0$, $s = 1$ and (7.3-24)

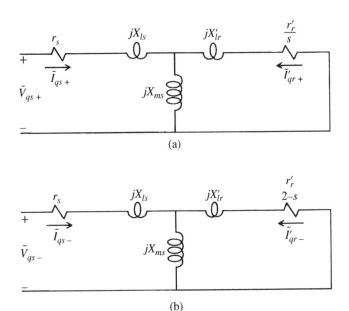

Figure 7.3-1 Equivalent-sequence circuits for unbalanced source voltages applied to a symmetrical two-phase induction motor. (a) Positive sequence; (b) negative sequence.

is zero. Hence, a steady-state pulsating torque does not exist at stall. Actually, the amplitude of the pulsating torque is zero at $\omega_r = 0$ regardless of the stator conditions. That is, an impedance may be in series with one of the stator windings or one winding may be opened-circuited. Regardless of the value of the sequence currents, (7.3-23) is zero when $s = 1$. The only requirement is that the rotor windings must be symmetrical. This is an interesting observation.

7.3.2 Unbalanced Stator Impedances

When an impedance is placed in series with the as winding of the stator, we can write

$$e_{ga} = i_{as}z(p) + v_{as} \qquad (7.3\text{-}25)$$

$$e_{gb} = v_{bs} \qquad (7.3\text{-}26)$$

where v_{as} and v_{bs} are the voltages across the stator phase windings and e_{ga} and e_{gb} are the source voltages which may be unbalanced. In (7.3-25), $z(p)$ is the operational notation of the impedance; for example, a series rL would be expressed $z(p) = r + pL$. The phasor equivalents of (7.3-25) and (7.3-26) are

$$\tilde{V}_{as} = \tilde{E}_{ga} - \tilde{I}_{as}Z \tag{7.3-27}$$

$$\tilde{V}_{bs} = \tilde{E}_{gb} \tag{7.3-28}$$

We can apply (7.2-9) to determine $\tilde{V}_{qs+}^s$ and $\tilde{V}_{qs-}^s$ as

$$\begin{bmatrix} \tilde{V}_{qs+}^s \\ \tilde{V}_{qs-}^s \end{bmatrix} = \frac{1}{2}\begin{bmatrix} 1 & j \\ 1 & -j \end{bmatrix}\begin{bmatrix} \tilde{E}_{ga} - \tilde{I}_{as}Z \\ \tilde{E}_{gb} \end{bmatrix} \tag{7.3-29}$$

which yields

$$\tilde{V}_{qs+}^s = \frac{1}{2}\left(\tilde{E}_{ga} + j\tilde{E}_{gb} - \tilde{I}_{as}Z\right) \tag{7.3-30}$$

$$\tilde{V}_{qs-}^s = \frac{1}{2}\left(\tilde{E}_{ga} - j\tilde{E}_{gb} - \tilde{I}_{as}Z\right) \tag{7.3-31}$$

Now,

$$\tilde{I}_{as} = \tilde{I}_{qs}^s = \tilde{I}_{qs+}^s + \tilde{I}_{qs-}^s \tag{7.3-32}$$

Substituting (7.3-32) into (7.3-30) and (7.3-31) for $\tilde{I}_{as}$ and then substituting the result into (7.3-6), and assuming the rotor windings are short-circuited, we obtain

$$\begin{bmatrix} \tilde{E}_1 \\ 0 \\ \tilde{E}_2 \\ 0 \end{bmatrix} = \begin{bmatrix} \frac{1}{2}Z + r_s + jX_{ss} & jX_{ms} & \frac{1}{2}Z & 0 \\ jX_{ms} & \dfrac{r_r'}{s} + jX_{rr}' & 0 & 0 \\ \frac{1}{2}Z & 0 & \frac{1}{2}Z + r_s + jX_{ss} & jX_{ms} \\ 0 & 0 & jX_{ms} & \dfrac{r_r'}{2-s} + jX_{rr}' \end{bmatrix}\begin{bmatrix} \tilde{I}_{qs+}^s \\ \tilde{I}_{qr+}^{\prime s} \\ \tilde{I}_{qs-}^s \\ \tilde{I}_{qr-}^{\prime s} \end{bmatrix} \tag{7.3-33}$$

where

$$\tilde{E}_1 = \frac{1}{2}\left(\tilde{E}_{ga} + j\tilde{E}_{gb}\right) \tag{7.3-34}$$

$$\tilde{E}_2 = \frac{1}{2}\left(\tilde{E}_{ga} - j\tilde{E}_{gb}\right) \tag{7.3-35}$$

If the impedance is a series capacitor, then $Z = -j(1/\omega_e C)$. Also, note that the positive- and negative-sequence voltage equations are now coupled. Although we could derive an equivalent circuit to portray these voltage equations, it is not worth the work. We can use (7.3-33) directly; however, a computer would be helpful. We shall work more with this equation when we analyze the symmetrical two-phase induction motor used as a single-phase motor.

7.3.3 Open-Circuited Stator Phase

For the analysis of an open-circuited stator phase, let us assume that i_{as} $\left(i_{qs}^s\right)$ is zero. Hence, from (2.4-13),

$$v_{qs}^s = p\lambda_{qs}^s \tag{7.3-36}$$

Now, since $i_{qs}^s = 0$, λ_{qs}^s may be expressed as

$$\lambda_{qs}^s = L_{ms}i_{qr}'^s \tag{7.3-37}$$

Since $v_{as} = v_{qs}^s$, we can write

$$v_{as} = L_{ms}pi_{qr}'^s \tag{7.3-38}$$

$$v_{bs} = e_{gb} \tag{7.3-39}$$

where e_{gb} is the source voltage. Now, in phasor form,

$$\tilde{V}_{as} = jX_{ms}\tilde{I}_{qr}'^s \tag{7.3-40}$$

$$\tilde{V}_{bs} = \tilde{E}_{gb} \tag{7.3-41}$$

Substituting (7.3-40) and (7.3-41) into (7.2-9), we obtain

$$\tilde{V}_{qs+}^s = \frac{1}{2}jX_{ms}\tilde{I}_{qr}'^s + \frac{1}{2}j\tilde{E}_{gb} \tag{7.3-42}$$

$$\tilde{V}_{qs-}^s = \frac{1}{2}jX_{ms}\tilde{I}_{qr}'^s - \frac{1}{2}j\tilde{E}_{gb} \tag{7.3-43}$$

From (7.3-1),

$$\tilde{I}_{qr}'^s = \tilde{I}_{qr+}'^s + \tilde{I}_{qr-}'^s \tag{7.3-44}$$

and, since $\tilde{I}_{qs}^s = 0$, (7.2-3) becomes

$$\tilde{I}_{qs-}^s = -\tilde{I}_{qs+}^s \tag{7.3-45}$$

Substituting (7.3-44) into (7.3-42) and (7.3-43), then substituting the result into (7.3-6), with (7.3-45) incorporated, we can write (rotor windings short-circuited)

$$\begin{bmatrix} \frac{1}{2}j\tilde{E}_{gb} \\ 0 \\ 0 \end{bmatrix} = \begin{bmatrix} r_s + jX_{ss} & j\frac{1}{2}X_{ms} & -j\frac{1}{2}X_{ms} \\ jX_{ms} & \frac{r_r'}{s} + jX_{rr}' & 0 \\ -jX_{ms} & 0 & \frac{r_r'}{2-s} + jX_{rr}' \end{bmatrix} \begin{bmatrix} \tilde{I}_{qs+}^s \\ \tilde{I}_{qr+}^{\prime s} \\ \tilde{I}_{qr-}^{\prime s} \end{bmatrix} \qquad (7.3\text{-}46)$$

From (7.3-40), the open-circuit voltage of the *as* winding is

$$\tilde{V}_{as} = jX_{ms}\left(\tilde{I}_{qr+}^{\prime s} + \tilde{I}_{qr-}^{\prime s}\right) \qquad (7.3\text{-}47)$$

where $\tilde{I}_{qr+}^{\prime s}$ and $\tilde{I}_{qr-}^{\prime s}$ are calculated from (7.3-46).

SP7.3-1. Determine the rotor speed at which the negative-sequence rotor currents $\tilde{I}_{qr-}^{\prime s}$ and $\tilde{I}_{dr-}^{\prime s}$ are zero for unbalanced applied stator voltages. $[\omega_r = -\omega_e]$

SP7.3-2. Assume that the steady-state T_e versus ω_r plot shown in Fig. 3.8-2 is for $\tilde{V}_{as} = j\tilde{V}_{bs}$. Plot the T_e versus ω_r for $\tilde{V}_{as} = -j\tilde{V}_{bs}$. [Inverted mirror image]

SP7.3-3. Determine the rotor speed at which $Z_+ = Z_-$ for a symmetrical induction motor. $[\omega_r = 0]$

SP7.3-4. Express Z_+ and Z_-. $[Z_+ = (3.8\text{-}14), Z_- = (3.8\text{-}14)$ with s replaced by $2-s]$

SP7.3-5. Express v_{bs} when $i_{bs} = 0$. $\left[v_{bs} = -L_{ms}pi_{dr}^{\prime s}\right]$

SP7.3-6. Throughout this work we have assumed ω_r constant for unbalanced operation. When could this assumption be in question? [low-frequency with unbalanced applied voltages]

7.4 Single-Phase and Capacitor-Start Induction Motors

7.4.1 Single-Phase Induction Motor

In Chapter 3, we talked briefly about single-phase induction motors. Although we will find that we must provide some means of starting the device, the single-phase induction motor has only one stator winding energized during normal operation. With this in mind, let us calculate the steady-state torque versus speed characteristics with voltage applied to only one stator winding of a symmetrical two-phase induction motor with the other winding open-circuited. Recall that we have already derived the voltage equations necessary to make these calculations. In particular, (7.3-46) can be used to determine the sequence currents with the *as* winding open-circuited and a single-phase voltage source connected to the *bs* winding. Once these calculations are made, the sequence currents may be substituted into

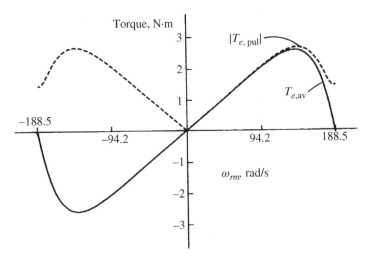

Figure 7.4-1 Steady-state torque-versus-speed characteristics for a single-phase induction motor $\left(\omega_{rm} = \left(\dfrac{2}{P} \right) \omega_r \right)$.

(7.3-17) to determine the average and pulsating components of the steady-state electromagnetic torque. The steady-state torque versus speed characteristics are shown in Fig. 7.4-1 for a symmetrical two-phase induction motor with rated voltage applied to one phase and the other phase open-circuited. The symmetrical two-phase induction machine is a four-pole $\dfrac{1}{4}$-hp 110-V 60-Hz motor with the following parameters: $r_s = 2.02\,\Omega$, $X_{ls} = 2.79\,\Omega$, $X_{ms} = 66.8\,\Omega$, $r_r' = 4.12\,\Omega$, and $X_{lr}' = 2.12\,\Omega$. The total inertia is $J = 1.46 \times 10^{-2}\,\mathrm{kg \cdot m^2}$.

The average steady-state electromagnetic torque $T_{e,\,\text{ave}} = T_{e+} - T_{e-}$ and the magnitude of the double-frequency component of the torque $|T_{e,\,\text{pul}}|$ are plotted in Fig. 7.4-1. There are at least two features worth mentioning. First, the plot of the average torque $T_{e,\,\text{ave}}$ for $\omega_{rm} < 0$ is the negative mirror image of that for $\omega_{rm} > 0$. Secondly, the plot of the pulsating torque $|T_{e,\,\text{pul}}|$ is symmetrical about the zero speed axis. Finally, we see verification of our earlier claim that the starting torque, is zero; $T_e = 0$ at $\omega_{rm} = 0$. Recall that $\omega_{rm} = (2/P)\omega_r$.

7.4.2 Capacitor-Start Induction Motor

As we know, the single-phase induction motor will not develop a starting torque since two equal and oppositely rotating air-gap mmf's are generated by a sinusoidal stator winding current. If, now, we take a two-phase symmetrical induction motor and apply a single-phase voltage across the two phases, the net torque at stall will be zero since the rotor currents will be instantaneously equal and the air-gap mmf will pulsate along an axis midway between the as and bs axes. Thus, two equal and oppositely rotating air-gap mmf's result. If, however, we cause the current in one of the phases to be different instantaneously from that in the other phase, a starting torque can be developed since this would cause one of the rotating air-gap mmf's to be larger than the other. One way of doing this is to place a capacitor in series with one of the stator windings of a two-phase symmetrical induction motor. This will cause the current in the phase with the series capacitor to lead the current in the other winding when the same voltage is applied to both.

We have already derived the equations necessary to calculate the component currents with an impedance in series with the as winding. In particular, (7.3-33) can be used to calculate the component currents with a capacitor in series with the as winding. If we set $Z = -j1/\omega_e C$ and let $\tilde{E}_{ga} = \tilde{E}_{gb}$, the single-phase source voltage, counterclockwise rotation will occur since $\tilde{I}_{as}$ will lead $\tilde{I}_{bs}$. Recall that, for the assumed positive direction of the magnetic axes and for a balanced two-phase set, we have $\tilde{I}_{as}$ leading $\tilde{I}_{bs}$ by 90° for counterclockwise rotation of the air-gap mmf.

Once the component currents are calculated, (7.3-17) can be used to determine the average steady-state electromagnetic torque $T_{e,\,ave}$ and the magnitude of the double-frequency component $|T_{e,\,pul}|$. These steady-state torque versus speed characteristics are shown in Fig. 7.4-2 for $C = 530.5\,\mu F$.

In capacitor-start single-phase induction motors, the winding with the series capacitor is disconnected from the source after the rotor has reached 60–80% of synchronous speed. This is generally accomplished by a centrifugal switching mechanism located inside the housing of the motor. Once the winding with the series capacitor is disconnected, the device then operates as a single-phase

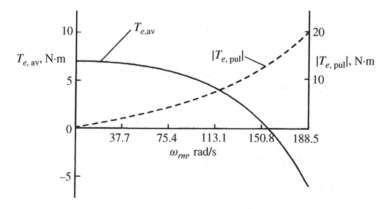

Figure 7.4-2 Steady-state torque-versus-speed characteristics with a capacitor in series with one winding of the two-phase induction machine.

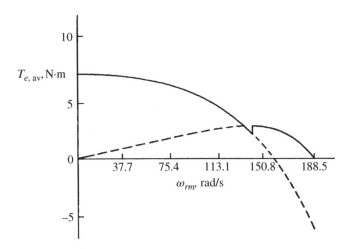

Figure 7.4-3 Average steady-state torque-versus-speed characteristics of a capacitor-start single-phase induction motor.

induction motor. In Fig. 7.4-3, the plot of average torque versus speed with a series capacitor in one phase (Fig. 7.4-2) is superimposed upon the plot of average torque versus speed with a single-phase winding (Fig. 7.4-1). The transition from capacitor-start to single-phase operation at 75% of synchronous speed is illustrated.

Although the capacitor-start single-phase induction motor is by far the most common type of single-phase induction motor, a capacitor-start capacitor-run induction motor is sometimes used. In this case, both phase windings are energized during normal operation. The value of the series capacitance is changed from the start value to the run value once the rotor reaches 60–80% of synchronous speed. This is accomplished using two capacitors connected in parallel with provision to open-circuit one of the parallel paths. The purpose of the run capacitor is to establish a leading current during normal loads, thereby increasing the torque capability over that which is possible with only one stator winding energized. Since two capacitors are needed, this device is somewhat more expensive and often the application does not justify this added cost.

SP7.4-1. Determine the frequency of the rotor currents when ω_{rm} is equal to synchronous speed in Fig. 7.4-1. [120 Hz]

SP7.4-2. For the torque-speed characteristics shown in Fig. 7.4-1, determine the approximate rotor speed at which the steady-state instantaneous torque first pulsates to a negative value. [$\omega_{rm} \cong 600$ r/min]

SP7.4-3. In Fig. 7.4-3, the device switches from capacitor-start to single-phase operation at a rotor speed of 75% of synchronous speed. Will the rotor accelerate faster or slower immediately following the switching? [Faster]

SP7.4-4. If the value of the capacitor was decreased, would you expect the starting torque to decrease or increase? Why? [Decrease, less leading component of current]

7.5 Dynamic and Steady-State Performance of a Capacitor-Start Single-Phase Induction Motor

The free-acceleration characteristics of the example capacitor-start single-phase induction motor are shown in Fig. 7.5-1. The variables v_{as}, i_{as}, v_{bs}, i_{bs}, v_c, T_e, and ω_{rm} are plotted. The voltage v_c is the instantaneous voltage across the capacitor

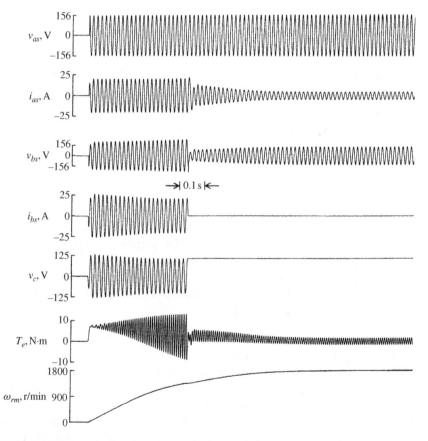

Figure 7.5-1 Free-acceleration characteristics of capacitor-start, single-phase induction motor.

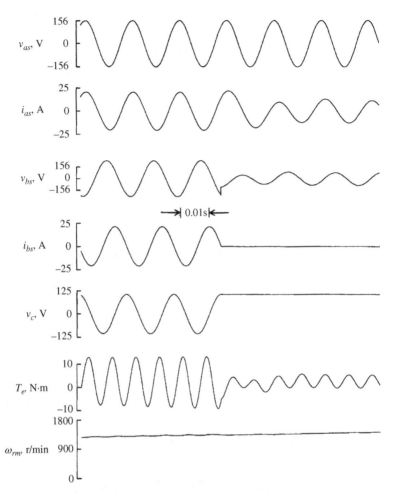

Figure 7.5-2 Expanded plot of Fig. 7.5-1 illustrating the disconnection of the capacitor in the *bs* winding.

which is connected in series with the *bs* winding. The machine variables are shown with an expanded scale in Fig. 7.5-2 to illustrate the switching out of the *bs* winding, which is disconnected from the source at a normal current zero once the rotor reaches 75% of synchronous speed. The voltage across the capacitor is shown to remain constant at its value when the *bs* winding is disconnected from the source. In practice, this voltage would slowly decay owing to leakage currents within the capacitor which are not considered in this analysis. The torque versus speed characteristics given in Fig. 7.5-3 are for the free acceleration shown in Fig. 7.5-1. The dynamic and steady-state characteristics following changes in load torque are

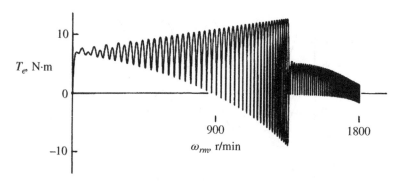

Figure 7.5-3 Torque-versus-speed characteristics for Fig. 7.5-1.

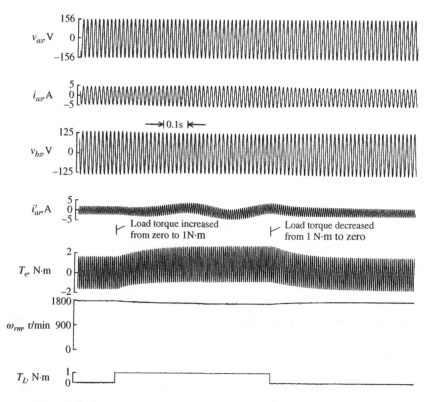

Figure 7.5-4 Step changes in load torque of single-phase induction motor.

illustrated in Fig. 7.5-4. Therein v_{as}, i_{as}, v_{bs}, (open-circuited), i'_{ar}, T_e, ω_{rm}, and T_L are plotted.

SP7.5-1. In Fig. 7.5-1, the capacitor is in series with the bs winding and the voltage applied to the as winding is $V_{as} = \sqrt{2}\,110\cos\omega_e t$. Calculate the steady-state stator currents I_{as} and I_{bs} at stall ($\omega_{rm} = 0$). Compare with the traces of i_{as} and i_{bs} in Fig 7.5-2. Neglect the magnetizing reactance X_{ms} in these calculations. $[I_{as} \cong 19.8\cos(377t - 38.6^\circ); I_{bs} \cong -25.3\cos(377t - 0.8^\circ)]$

SP7.5-2. Determine the frequency of I'^s_{qr} and I'^s_{dr} for the loaded condition ($T_L = 1$ N $\cdot$ m) in Fig. 7.5-4. [60 Hz]

7.6 Split-Phase Induction Motor

Although we have considered only the symmetrical two-phase induction motor as a single-phase induction motor, the split-phase induction motor is often used [3, 4]. It is an unsymmetrical two-phase induction machine; that is, the stator windings are different. The main or run winding remains energized during normal operation while the start or auxiliary winding is switched out after the rotor reaches 60–80% of synchronous speed. The r to X ratio of the run winding would be much the same as that of the stator windings of a two-phase symmetrical machine; however, the start winding has a higher r to X ratio. Hence, with the same voltage applied to the start and run windings, the current flowing in the start winding would lead the current flowing in the run winding. We see the logic behind all of this. Rather than using only a capacitor to shift the phase of one of the winding currents in order to develop a starting torque, the machine is designed with different stator windings so that one current leads the other due to the difference in the winding impedances. Depending on the application and the design of the machine, a capacitor may or may not be used in series with the start winding.

We will not analyze the split-phase induction machine. The analysis is rather involved since the mutual inductances between the rotor windings and the run winding are different from those between the rotor windings and the start winding. Actually we have established the main operating characteristics of single-phase induction motors with the least amount of effort by considering the symmetrical two-phase machine. If, however, one wishes to consider the split-phase device in more detail, this analysis is given in [3].

7.7 Problems

1 Calculate $\tilde{F}^s_{qs+}$ and $\tilde{F}^s_{qs-}$ for the following sets using (7.2-9).

(a) $\tilde{F}_{as} = 10\underline{/30°}, \tilde{F}_{bs} = 30\underline{/-60°}$.

(b) $\tilde{F}_{as} = 10\underline{/0°}, \tilde{F}_{bs} = 0$.

(c) $\tilde{F}_{as} = \cos(\omega_e t + 45°), \tilde{F}_{bs} = \cos(\omega_e t - 45°)$.

2 Start with (7.3-5) and derive (7.3-6).

3 Derive (7.3-16).

4 Show that (7.3-16) and (7.3-17) are equivalent.

5 Express (7.3-46) with $\tilde{I}^s_{qr+}$ and $\tilde{I}^s_{qr-}$ eliminated.

6 The equivalent circuit for steady-state operation of an induction motor with only one stator winding is shown in Fig. 7.7-1. Show that this equivalent circuit is the same as that given by (7.3-46).

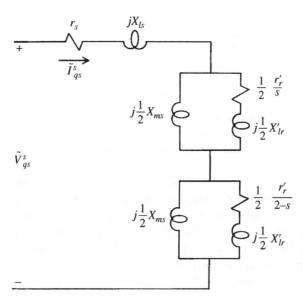

Figure 7.7-1 Equivalent circuit for single-phase stator winding.

References

1 C. L. Fortescue, "Method of Symmetrical Co-ordinates Applied to the Solution of Polyphase Networks," *AIEE Trans.*, vol. 37, 1918, pp. 627–1115.

2 P. C. Krause, "Method of Symmetrical Components Derived by Reference Frame Theory," *IEEE Trans., Power Appar. Syst.*, vol. 64, June 1985, pp. 1492–1499.

3 P. C. Krause, "Simulation of Unsymmetrical Two-Phase Induction Machines," *IEEE Trans. Power Appar. Syst.*, vol. 84, November 1965, pp. 1025–1037.

4 P. C. Krause, *Analysis of Electrical Machinery*, New York; McGraw-Hill Book Company, 1986.

8

Stepper Motors

8.1 Introduction

Stepper motors are electromechanical motion devices which are used primarily to convert information in digital form to mechanical motion. Although stepper motors were used as early as the 1920s, their use has skyrocketed with the advent of the digital computer. Whenever stepping from one position to another is required, whether the application is industrial, military, or medical, the stepper motor is often the motor of choice. Stepper motors come in various sizes and shapes but most fall into two types – the variable-reluctance stepper motor and the permanent-magnet stepper motor. Both types are considered in this chapter. We shall find that the operating principle of the variable-reluctance stepper motor is much the same as that of the salient-pole (reluctance) machine, and the permanent-magnet stepper motor is similar in principle to the permanent-magnet synchronous or ac machine.

8.2 Basic Configurations of Multistack Variable-Reluctance Stepper Motors

There are two general types of variable-reluctance stepper motors: single- and multistack. As a first approximation, the behavior of both types may be described from similar equations. Actually, the principle of operation of variable-reluctance stepper motors is similar to the reluctance torque which we considered in Chapter 4; only the mode of operation differs. There are, however, some new terms to define, and it is necessary for us to extend some of our previous definitions to fit the stepper motor. First, we will look at the multistack device in some detail, followed by a brief discussion of the single-stack variable-reluctance stepper motor.

Introduction to Modern Analysis of Electric Machines and Drives, First Edition.
Paul C. Krause and Thomas C. Krause.
© 2023 The Institute of Electrical and Electronics Engineers, Inc.
Published 2023 by John Wiley & Sons, Inc.

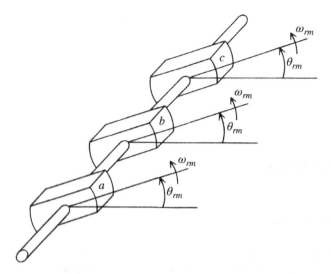

Figure 8.2-1 Rotor of an elementary two-pole, three-stack, variable-reluctance stepper motor.

In its most basic form, the multistack variable-reluctance stepper motor consists of three or more single-phase reluctance motors on a common shaft with their stator magnetic axes displaced from each other. The rotor of an elementary three-stack device is shown in Fig. 8.2-1. It has three cascaded two-pole rotors with a minimum-reluctance path of each aligned at the angular displacement θ_{rm}. In stepper motor language, each of the two-pole rotors is said to have two teeth. Now, visualize that each of these rotors has its own, separate, single-phase stator with the magnetic axes of the stators displaced from each other. In Fig. 8.2-1, we have labeled the individual rotors a, b, and c. The corresponding stators are shown in Fig. 8.2-2; the stator with the as winding is associated with the a rotor, the bs winding with the b rotor, etc. There are several things to note. First, we see that each of the single-phase stators has two poles, with the stator winding wound around both poles. In particular, positive current flows into as_1 and out as_1', which is then connected to as_2 so that positive current flows into as_2 and out as_2'. Although we have shown only one circle for $as_1, \ldots as_2'$, we realize that each would represent several turns, and that the number of turns from as_1 to as_1' (indicated by $N_s/2$ in Fig. 8.2-2) is the same as from as_2 to as_2'. Let us note one more thing; heretofore, we have referenced θ_{rm} (or θ_r) from the as axis to the maximum-reluctance path of a salient-pole rotor as shown in Fig. 4.3-1. In Fig. 8.2-2, θ_{rm} is referenced to the minimum-reluctance path of the rotor. Since this is more or less standard in stepper motor analysis, we will deviate from the convention we have established for synchronous machines.

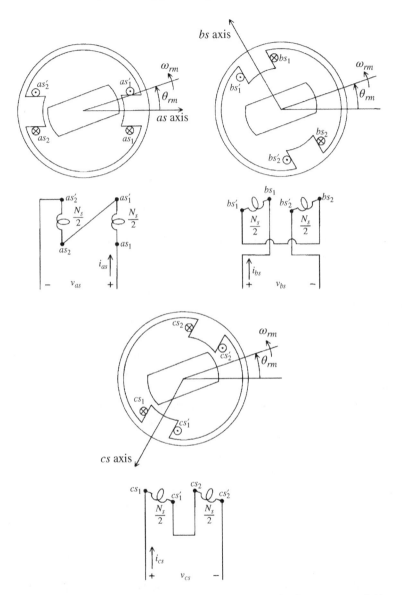

Figure 8.2-2 Stator configuration for an elementary two-pole, three-stator variable-reluctance stepper motor.

Each stack is often called a phase. In other words, a three-stack machine is a three-phase machine. This nomenclature can be misleading since we generally think of a three-phase ac device when we hear the words three-phase machine. We will find that a stepper motor is a discrete device, operated by switching a dc voltage from one stator winding to the other. Although more than three stacks (phases) may be used, three-stack variable-reluctance stepper motors are quite common. Our previous meaning of phase must be changed somewhat to accommodate the stepper motor.

Before writing any equations, let us see if we can gain some insight in regard to the operation of this device. To start, let the *bs* and *cs* windings in Fig. 8.2-2 be open-circuited, and let us apply a dc voltage to the *as* winding, whereupon we will assume that a constant i_{as} is immediately established. Now, since the magnetic systems of the three single-phase stators are separate, flux set up by one winding does not link the other windings. Hence, with only the *as* winding energized, flux exists only in the *as* axis. The minimum-reluctance path of the *a* part of the rotor (see Fig. 8.2-2) will align with the *as* axis. That is, at equilibrium with zero load torque, θ_{rm} in all parts of Fig. 8.2-2 is the same, either zero or 180°; let us say it is zero to make our discussion easier. (What would the rotor do if we could instantaneously reverse the direction of i_{as}?)

Stepper motors are used to convert digital or discrete information into a change in angular position. Let us see how positioning (stepping) is achieved. For this, let us instantaneously deenergize the *as* winding and immediately establish a direct current in the *bs* winding. The minimum reluctance path of the rotor will align itself with the *bs* axis. To do this, the rotor would rotate clockwise from $\theta_{rm} = 0$ to $\theta_{rm} = -60°$. Note that by advancing the mmf from the positive *as* axis to the positive *bs* axis, 120° counterclockwise, we have caused a 60° clockwise rotation of the rotor. There must be something wrong here. We recall from our work with rotating magnetic fields that, with the magnetic axes as shown in Fig. 8.2-2, an *abc* sequence of balanced sinusoidal currents will yield operation at synchronous speed with the rotor rotating counterclockwise. Therefore, it would seem that rotating the air-gap mmf from the positive *as* axis to the positive *bs* axis would cause rotation in the counterclockwise direction. In the case of variable-reluctance stepper motors, we will find that the direction of stepping can be either in the same or opposite direction of the rotation of the air-gap mmf depending upon the number of phases (stacks), the number of poles created by the stator windings, and the number of rotor teeth.

If, instead of energizing the *bs* winding, we energize the *cs* winding in Fig. 8.2-2, the rotor would have stepped counterclockwise from $\theta_{rm} = 0$ to $\theta_{rm} = 60°$. Thus, applying a dc voltage separately in the sequences *as, bs, cs, as, ...* produces 60° steps in the clockwise direction, whereas the sequence *as, cs, bs, as, ...* produces 60° steps in the counterclockwise direction. We need at least three stacks to achieve rotation (stepping) in both directions.

Before defining some stepper motor terms, let us think of one more thing. What if we energized the *as* and *bs* windings with the same current? That is, assume that initially the *as* winding is energized with $\theta_{rm} = 0$ and the *bs* winding is energized without deenergizing the *as* winding. What happens? Well, the rotor rotates clockwise from $\theta_{rm} = 0$ to $\theta_{rm} = -30°$. We have reduced our step length by one half. This is referred to as *half-step operation*. What would happen if all three stacks were excited with a positive voltage?

It is time to define terms. Let *RT* denote the number of rotor teeth per stack and *ST* the number of stator teeth per stack. The elementary device shown in Figs. 8.2-1 and 8.2-2 has two poles, two rotor teeth, and two stator teeth per stack; thus, $RT = ST = 2$. In fact, *RT* (rotor teeth per stack) always equals *ST* (stator teeth per stack) in a multistack variable-reluctance stepper motor. The number of stacks is denoted as *N*; here $N = 3$. Now, the tooth pitch, which we will denote as *TP*, is the angular displacement between rotor teeth. In this case, $TP = 180°$. We can write

$$TP = \frac{2\pi}{RT} \tag{8.2-1}$$

We have one more term to define; the step length, denoted as *SL*. It is the angular rotation of the rotor as we change the excitation (dc voltage) from one phase to the other. In this case, the step length is 60°, $SL = 60°$. If we energize each stack separately, then going from *as* to *bs* to *cs* back to *as* causes the rotor to rotate one tooth pitch. In other words, the number of stacks (phases) times the step length is a tooth pitch. That is,

$$TP = N\, SL \tag{8.2-2}$$

We can substitute (8.2-1) into (8.2-2) and obtain

$$SL = \frac{TP}{N} = \frac{2\pi}{RT\, N} \tag{8.2-3}$$

We shall find use for all of these new terms as we go along.

Although the elementary device shown in Figs. 8.2-1 and 8.2-2 offers a good starting point in our analysis of stepper motors, it has limited application owing to its large step length. Let us consider the four-pole three-stack variable-reluctance stepper motor with four rotor teeth, as illustrated in Fig. 8.2-3. Here, $RT = 4$ and $N = 3$; therefore, from (8.2-1), the tooth pitch is $TP = 2\pi/RT = 90°$. From (8.2-2), the step length is $SL = TP/N = 30°$ and an *as, bs, cs, as, ...* sequence produces 30° steps in the clockwise direction.

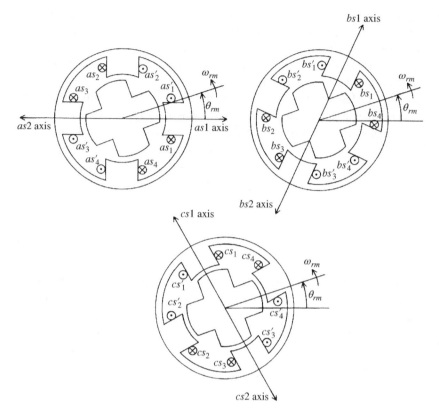

Figure 8.2-3 Four-pole, three-stack, variable-reluctance stepper motor with four rotor teeth.

The device shown in Fig. 8.2-4 is a four-pole three-stack variable-reluctance stepper motor with eight rotor teeth. In this case, $RT = 8$ and $N = 3$, thus, $TP = 45°$ and $SL = 15°$. However, in this device an *as, bs, cs, as,* ... sequence produces $15°$ steps in the counterclockwise direction. The pattern is clear; by increasing the number of stator and rotor teeth, we reduce the step length. The step lengths of multistack variable-reluctance stepping motors typically range from 2 to $15°$.

There appears to be an inconsistency in Fig. 8.2-4. In particular, θ_{rm} is referenced from the *as* axis to a position between rotor teeth. Earlier in this section, we established that, in the case of stepper motors, we would reference θ_{rm} from the *as* axis to the minimum-reluctance path of the rotor, whereupon the reluctance of the magnetic system associated with the *as* winding would be minimum when $\theta_{rm} = 0$.

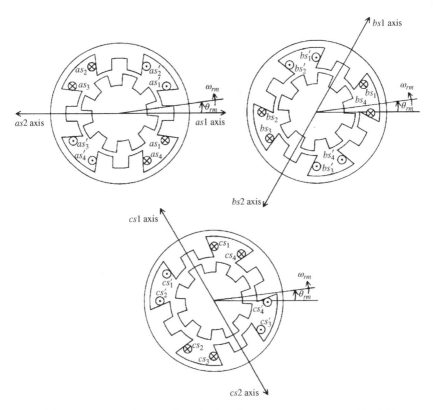

Figure 8.2-4 Four-pole, three-stack, variable-reluctance stepper motor with eight rotor teeth.

At first glance it appears that we have violated this stepper motor convention. However, when θ_{rm} is zero in Fig. 8.2-4, the reluctance of the magnetic system associated with the *as* winding is minimum. Hence, we must reference θ_{rm} from a position between rotor teeth to maintain the convention which we established earlier in this section. A cutaway view of a four-pole three-stack variable-reluctance stepper motor with 16 rotor teeth is shown in Fig. 8.2-5.

SP8.2-1. Calculate the step length for an eight-pole three-stack variable-reluctance stepper motor with 16 rotor teeth. [$SL = 7.5°$]

SP8.2-2. Consider the two-pole two-phase reluctance motor. Calculate (*a*) TP, (*b*) SL, and (*c*) determine the direction of rotation when a dc voltage is switched from the *as* winding to the *bs* winding. [(*a*) $TP = 180°$; (*b*) $SL = 90°$; (*c*) either ccw or cw]

Figure 8.2-5 Cutaway view of four-point, three-stack, variable-reluctance stepper motor with 16 rotor teeth. *Source:* Courtesy of Warner Electric.

8.3 Equations for Multistack Variable-Reluctance Stepper Motors

The voltage equations for a three-stack variable-reluctance stepper motor may be written as

$$v_{as} = r_s i_{as} + \frac{d\lambda_{as}}{dt} \tag{8.3-1}$$

$$v_{bs} = r_s i_{bs} + \frac{d\lambda_{bs}}{dt} \tag{8.3-2}$$

$$v_{cs} = r_s i_{cs} + \frac{d\lambda_{cs}}{dt} \tag{8.3-3}$$

In matrix form,

$$\mathbf{v}_{abcs} = \mathbf{r}_s \mathbf{i}_{abcs} + p\boldsymbol{\lambda}_{abcs} \tag{8.3-4}$$

where p is the operator d/dt and, for voltages, currents, and flux linkages

$$(\mathbf{f}_{abcs})^T = [f_{as} \quad f_{bs} \quad f_{cs}] \tag{8.3-5}$$

with

$$\mathbf{r}_s = \begin{bmatrix} r_s & 0 & 0 \\ 0 & r_s & 0 \\ 0 & 0 & r_s \end{bmatrix} \tag{8.3-6}$$

Since magnetic coupling does not exist between phases, we can write the flux linkages as

$$\begin{bmatrix} \lambda_{as} \\ \lambda_{bs} \\ \lambda_{cs} \end{bmatrix} = \begin{bmatrix} L_{asas} & 0 & 0 \\ 0 & L_{bsbs} & 0 \\ 0 & 0 & L_{cscs} \end{bmatrix} \begin{bmatrix} i_{as} \\ i_{bs} \\ i_{cs} \end{bmatrix} \tag{8.3-7}$$

For the purpose of expressing the self-inductances L_{asas}, L_{bsbs}, and L_{cscs}, let us first consider the elementary two-pole device illustrated in Fig. 8.2-2. We can write as a first approximation,

$$L_{asas} = L_{ls} + L_A + L_B \cos 2\theta_{rm} \tag{8.3-8}$$

$$L_{bsbs} = L_{ls} + L_A + L_B \cos 2\left(\theta_{rm} - \frac{2}{3}\pi\right) \tag{8.3-9}$$

$$L_{cscs} = L_{ls} + L_A + L_B \cos 2\left(\theta_{rm} - \frac{4}{3}\pi\right) \tag{8.3-10}$$

We are aware that L_{ls} is the leakage inductance whereas L_A and L_B are constants with $L_A > L_B$. The rotor displacement is expressed as

$$\theta_{rm} = \omega_{rm}t + \theta_{rm}(0) \tag{8.3-11}$$

We will use θ_{rm}, the actual rotor displacement, rather than θ_r, the electrical angular displacement. Although θ_{rm} and θ_r are related, $\theta_r = (P/2)\theta_{rm}$, where P is the number of poles, we will find it more convenient to use θ_{rm} in the analysis of stepper motors. We see that (8.3-8) is similar to (4.2-37), with θ_{rm} referenced to the minimum reluctance path of the rotor. Equation (8.3-7) is easily developed once we realize that the self-inductance of the bs winding is the same as that of the as winding. However, since θ_{rm} is referenced from the as axis, the angular displacement to the bs axis from the as axis must be subtracted from θ_{rm} so that, when $\theta_{rm} = \frac{2}{3}\pi$, the argument of (8.3-7) is zero and (8.3-7) with $\theta_{rm} = \frac{2}{3}\pi$ becomes the same as (8.3-8) with $\theta_{rm} = 0$. Following this same line of reasoning, we would determine that the angular displacement of (8.3-10) is $-\frac{4}{3}\pi$. However, since $\cos 2\left(\theta_{rm} - \frac{4}{3}\pi\right) = \cos 2\left(\theta_{rm} + \frac{2}{3}\pi\right)$, we can use $\frac{2}{3}\pi$ as the angular displacement for L_{cscs}. It is obvious that we can express (8.3-8) through (8.3-10) in various forms. Later, we will find it advantageous to express the argument of (8.3-8) and (8.3-10) in terms of step length.

The self-inductances of the four-pole three-stack variable-reluctance device with four rotor teeth shown in Fig. 8.2-3 can be approximated as

$$L_{asas} = L_{ls} + L_A + L_B \cos 4\theta_{rm} \tag{8.3-12}$$

$$L_{bsbs} = L_{ls} + L_A + L_B \cos 4\left(\theta_{rm} - \frac{1}{3}\pi\right) \qquad (8.3\text{-}13)$$

$$L_{cscs} = L_{ls} + L_A + L_B \cos 4\left(\theta_{rm} - \frac{2}{3}\pi\right) \qquad (8.3\text{-}14)$$

Although we are using the same L_{ls}, L_A, and L_B to denote constants, we realize that these are not equal for the various machines.

For the four-pole three-stack variable-reluctance stepper motor with eight rotor teeth shown in Fig. 8.2-4, we can approximate the self-inductances as

$$L_{asas} = L_{ls} + L_A + L_B \cos 8\theta_{rm} \qquad (8.3\text{-}15)$$

$$L_{bsbs} = L_{ls} + L_A + L_B \cos 8\left(\theta_{rm} - \frac{1}{3}\pi\right) \qquad (8.3\text{-}16)$$

$$L_{cscs} = L_{ls} + L_A + L_B \cos 8\left(\theta_{rm} - \frac{2}{3}\pi\right) \qquad (8.3\text{-}17)$$

By adding or subtracting multiples of 2π from the arguments, the above self-inductances may be expressed in terms of SL. In particular, for the devices shown in Figs. 8.2-2 and 8.2-3, where we have previously noted that a counterclockwise rotation of the stator' mmf and stepping are in opposite directions, cw, the inductances may be expressed as

$$L_{asas} = L_{ls} + L_A + L_B \cos\left(RT\,\theta_{rm}\right) \qquad (8.3\text{-}18)$$

$$L_{bsbs} = L_{ls} + L_A + L_B \cos\left[RT(\theta_{rm} + SL)\right] \qquad (8.3\text{-}19)$$

$$L_{cscs} = L_{ls} + L_A + L_B \cos\left[RT(\theta_{rm} - SL)\right] \qquad (8.3\text{-}20)$$

For the device shown in Fig. 8.2-4, where rotation of the stator mmf and stepping are in the same direction, ccw, the self-inductances may be expressed as

$$L_{asas} = L_{ls} + L_A + L_B \cos\left(RT\,\theta_{rm}\right) \qquad (8.3\text{-}21)$$

$$L_{bsbs} = L_{ls} + L_A + L_B \cos\left[RT(\theta_{rm} - SL)\right] \qquad (8.3\text{-}22)$$

$$L_{cscs} = L_{ls} + L_A + L_B \cos\left[RT(\theta_{rm} + SL)\right] \qquad (8.3\text{-}23)$$

An expression for the electromagnetic torque may be written,

$$T_e = \frac{\partial W_c(\mathbf{i}, \theta_{rm})}{\partial \theta_{rm}} \qquad (8.3\text{-}24)$$

Since we are assuming a linear magnetic system, the field energy and coenergy are equal. Thus, since the mutual inductances are zero,

$$W_c = \frac{1}{2}L_{asas}i_{as}^2 + \frac{1}{2}L_{bsbs}i_{bs}^2 + \frac{1}{2}L_{cscs}i_{cs}^2 \qquad (8.3\text{-}25)$$

Substituting the self-inductances given by (8.3-18) through (8.3-20) into (8.3-25) and taking the partial derivative with respect to θ_{rm} yields

$$T_e = -\frac{RT}{2}L_B\{i_{as}^2\sin(RT\,\theta_{rm}) + i_{bs}^2\sin[RT(\theta_{rm}+SL)] + i_{cs}^2\sin[RT\,(\theta_{rm}-SL)]\}$$

(8.3-26)

An alternate form of (8.3-26) using the tooth pitch TP is

$$
\begin{aligned}
T_e = -\frac{RT}{2}L_B\Big\{ & i_{as}^2\sin\left(\frac{2\pi}{TP}\theta_{rm}\right) + i_{bs}^2\sin\left[\frac{2\pi}{TP}\left(\theta_{rm}+\frac{TP}{3}\right)\right] \\
& + i_{cs}^2\sin\left[\frac{2\pi}{TP}\left(\theta_{rm}-\frac{TP}{3}\right)\right]\Big\}
\end{aligned}
$$

(8.3-27)

It is important to note that (8.3-26) and (8.3-27) are written for rotation of the stator mmf and stepping of the rotor in opposite directions. For stepping in the same direction, the sign preceding both SL's in (8.3-26) and both $(TP/3)$'s in (8.3-26) must be changed. Note also that the magnitude of the torque is proportional to the number of rotor teeth RT.

The torque and rotor angular position are related as

$$T_e = J\frac{d^2\theta_{rm}}{dt^2} + B_m\frac{d\theta_{rm}}{dt} + T_L$$

(8.3-28)

where J is the total inertia in $\text{kg}\cdot\text{m}^2$ and B_m is a damping coefficient associated with the mechanical rotational system in $\text{N}\cdot\text{m}\cdot\text{s}$. The electromagnetic torque T_e is positive in the counterclockwise direction (positive direction of θ_{rm}) whereas the load torque T_L is positive in the clockwise direction.

SP8.3-1. The stator currents of a three-stack variable-reluctance machine are $i_{as}=I$, $i_{bs}=-I$, and $i_{cs}=0$. Determine the no-load rotor position. $[\theta_{rm}=-TP/6]$

SP8.3-2. Repeat SP8.3-1 with $i_{as}=i_{bs}=i_{cs}$. $[T_e$ is zero for all values of $\theta_{rm}.]$

8.4 Operating Characteristics of Multistack Variable-Reluctance Stepper Motors

It is instructive to take a little closer look at the operating characteristics of a multistack variable-reluctance stepper motor from the standpoint of idealized, pseudo steady-state conditions. For this purpose, let us consider the expression for torque given by (8.3-27) for a three-stack motor with opposite directions of rotation of the stator mmf and stepping. In particular,

$$T_e = -\frac{RT}{2} L_B \left\{ i_{as}^2 \sin\left(\frac{2\pi}{TP}\theta_{rm}\right) + i_{bs}^2 \sin\left[\frac{2\pi}{TP}\left(\theta_{rm} + \frac{TP}{3}\right)\right]\right.$$
$$\left. + i_{cs}^2 \sin\left[\frac{2\pi}{TP}\left(\theta_{rm} - \frac{TP}{3}\right)\right]\right\}$$

(8.4-1)

In Fig. 8.4-1, the three terms of (8.4-1) are plotted separately for equal, constant (steady-state) currents. Let us assume that there is no load torque, $T_L = 0$, and $i_{as} = I$ while i_{bs} and i_{cs} are zero. Only the first term of (8.4-1) is present; that is, only the steady-state torque due to i_{as} exists. The stable steady-state rotor position would be at $\theta_{rm} = 0$ denoted as point 1 on Fig. 8.4-1. Now, let us assume that i_{as} is instantaneously decreased from I to zero while i_{bs} is increased from zero to I. Hence, the steady-state torque plot due to i_{as} would instantaneously disappear from Fig. 8.4-1 and the torque due to i_{bs} would immediately appear. Now, we know that this cannot happen in practice since there would be electrical transients involved, but we are neglecting all transients in this discussion. Since, the torque at point 2 is negative, the rotor will rotate in the clockwise, $-\theta_{rm}$, direction. We will then proceed along the i_{bs} torque plot until we have reached point 3. Note we have moved one step length in the clockwise direction. If, instead of energizing the bs winding after deenergizing the as winding, we energized the cs winding, then the torque at point

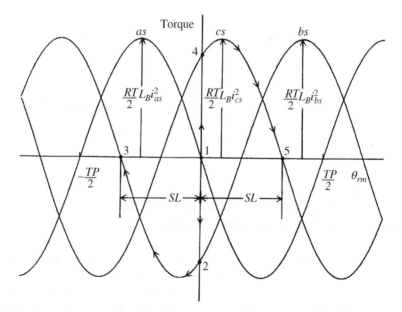

Figure 8.4-1 Stepping operation of a three-stack, variable-reluctance stepper motor without load torque – steady-state torque-angle plots.

4 would appear. This is a positive T_e so the rotor will rotate in the counterclockwise direction, $+\theta_{rm}$, and we will ride along the torque angle plot to point 5. Please realize that not only are we neglecting the electrical transients in this discussion but we are also neglecting the mechanical transients. Normally, there would be a damped oscillation about the new operating point.

Half-step operation is depicted in Fig. 8.4-2. To explain this, let us again start at point 1 where $T_L = 0$ and only the *as* winding is energized ($i_{as} = I$). Instantaneously, the *bs* winding is energized and $i_{bs} = I$. Now, both i_{as} and i_{bs} are I and only the *as* + *bs* torque plot, shown in Fig. 8.4-2, exists. Immediately, the torque at point 2 appears and the rotor starts to rotate in the clockwise, $-\theta_{rm}$, direction coming to rest at point 3. The rotor has moved $SL/2$ clockwise.

Stepping action with a load torque is shown in Fig. 8.4-3. Assume that initial operation is at point 1 with $i_{as} = I$ and $i_{bs} = i_{cs} = 0$. Recall that T_e is positive in the counterclockwise direction while T_L is positive in the clockwise direction, and stable operation occurs when $T_e = T_L$. Thus, at point 1, $T_e = T_L$. The *as* winding is deenergized while the *bs* winding is energized. Immediately, the negative T_e at point 2 appears and the rotor will move clockwise to point 3. If the cs winding is energized rather than the *bs* winding, the torque at point 4 would appear and the rotor would move counterclockwise to point 5. Note that the step length is still the same in both directions. However, the rotor will move more rapidly in the

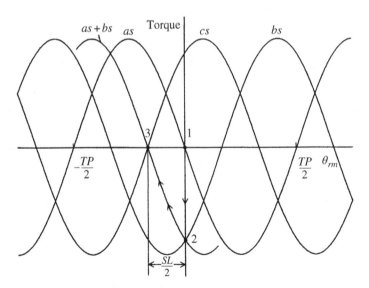

Figure 8.4-2 Half-step operation of a three-stack, variable-reluctance stepper motor – steady-state torque-angle plots.

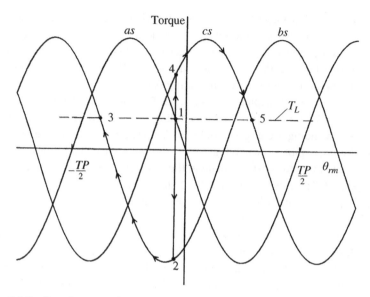

Figure 8.4-3 Stepping operation of a three-stack, variable-reluctance stepper motor with load torque – steady-state torque-angle plots.

clockwise direction than in the counterclockwise direction since the load torque is in the clockwise direction. In other words, there is a larger torque to accelerate the rotor in the clockwise direction than in the counterclockwise direction.

The plots of i_{as}, i_{bs}, i_{cs}, and θ_{rm} shown in Fig. 8.4-4 allow us to view stepping operation from another standpoint. Initially, there is no load torque and $i_{as} = I$. The current i_{as} is stepped off and i_{bs} is stepped on. The rotor rotates clockwise to $\theta_{rm} = -SL$. Here, we have indicated the presence of a damped mechanical oscillation which was not shown in the steady-state torque angle plots. Next, i_{bs} is switched off and i_{as} is switched back on. The rotor ends up back at $\theta_{rm} = 0$. Next, we see half-step operation; i_{as} remains at I while i_{cs} is switched to I. The rotor advances to $\frac{1}{2}SL$. When i_{as} is switched to zero, the rotor again advances by $\frac{1}{2}SL$ to $\theta_{rm} = SL$.

SP8.4-1. In Fig. 8.4-3, the load torque is such that the initial operating point with $i_{as} = I$ and $i_{bs} = i_{cs} = 0$ is at $\theta_{rm} = -TP/8$. The current in the as winding is switched to zero and the current in the cs winding is switched to I. Determine the final value of θ_{rm}. Which direction will the rotor rotate? [$\theta_{rm} = -TP/8 - 2SL$; cw]

SP8.4-2. It is desirable to step from $\theta_{rm} = 0$ to $\theta_{rm} = -SL/3$ for the device shown in Fig. 8.2-3. Assume that we have the facility to control the winding currents. Let $i_{as} = I$; determine i_{bs}. [$i_{bs}=0.81I$]

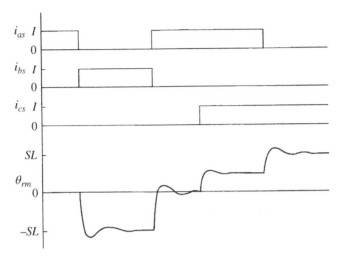

Figure 8.4-4 Stepping operation depicting θ_{rm} versus time – no load torque.

8.5 Single-Stack Variable-Reluctance Stepper Motors

As its name suggests, the single-stack variable-reluctance stepper motor has only one stack and all stator phases are arranged on this single stack. A three-phase single-stack variable-reluctance stepper motor is shown in Fig. 8.5-1. Here, it appears that we have taken the three two-pole single-phase stators shown in Fig. 8.2-2 and squeezed them into one stack. The magnetic axes of the stator windings are displaced 120° as in the case of the three-phase machines considered in earlier chapters; however, the stepper motor generally has stator teeth or poles which protrude rather than a circular inner stator surface.

Recall that in the case of the multistack variable-reluctance motor, the number of rotor and stator teeth per stack is the same. In the case of the single-stack stepper motor, the number of rotor teeth per stack, RT, is never equal to the number of stator teeth per stack, ST. If, for example, the rotor shown in Fig. 8.5-1 had the same number of teeth as the stator (6), then, when two diagonally opposite rotor teeth are aligned with two diagonally opposite stator teeth, all diagonally opposite rotor teeth would be aligned with diagonally opposite stator teeth and stepping action could not occur. The equations which we derived for the tooth pitch TP and step length SL for the multistack variable-reluctance stepper motor also apply for the single-stack stepper motor. For the two-pole three-phase stepper motor shown in Fig. 8.5-1, RT = 4 and, thus, $TP = 2\pi/RT = 90°$ and $SL = TP/N = 30°$. Note that the sequence as, bs, cs, as, ... produces a counterclockwise stepping of the rotor.

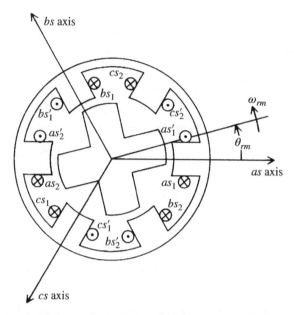

Figure 8.5-1 Two-pole, three-phase, single-stack, variable-reluctance stepper motor with six stator teeth and four rotor teeth.

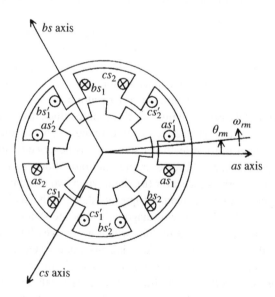

Figure 8.5-2 Two-pole, three-phase, single-stack, variable-reluctance stepper motor with six stator teeth and eight rotor teeth.

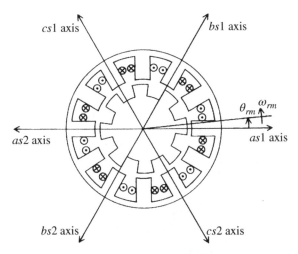

Figure 8.5-3 Four-pole, three-phase, single-stack, variable-reluctance stepper motor with 12 stator teeth and eight rotor teeth.

Two other types of three-phase single-stack variable-reluctance stepper motors are shown in Figs. 8.5-2 and 8.5-3. The two-pole device shown in Fig. 8.5-2 has six stator teeth and eight rotor teeth, $TP = 45°$ and $SL = 15°$, and an *as, bs, cs, as, ...* sequence produces a clockwise stepping of the rotor. For the four-pole three-phase device shown in Fig. 8.5-3, $ST = 12$ and $RT = 8$. Thus, $TP = 40°$, $SL = 15°$. The step length is the same as for the stator with six teeth (Fig. 8.5-2); however, counterclockwise stepping of the rotor occurs with the sequence *as, bs, cs, as, ...* In Fig. 8.5-3, the labeling of the coil sides of the windings is omitted because of lack of space.

The expressions given for the self-inductances of the three-stack (phase) variable-reluctance stepper motor, (8.3-18) through (8.3-23), also apply to the three-phase single-stack variable-reluctance stepper motor. Therefore, it would appear that the operation of the single-stack and multistack variable-reluctance stepper motors may be described by the same set of equations. Although this perception is essentially valid from an idealized point of view, it is not valid in the practical world. We see from Figs. 8.5-1 through 8.5-3 that the stator windings share the same magnetic system. Hence, there is a possibility of mutual coupling between stator phases. For the purposes of discussion, let us consider Fig. 8.5-4, which is Fig. 8.5-1 with $\theta_{rm} = 0$. The dashed lines shown therein depict the flux linking the *bs* winding due to positive current flowing in the *as* winding. If we assume that the reluctance of the iron is small so that it can be neglected, the flux linkages cancel, whereupon mutual coupling would not exist between stator phases. From an idealized standpoint, this is a valid line of reasoning; from a practical standpoint it is not.

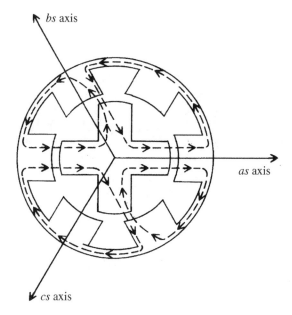

Figure 8.5-4 Two-pole, three-phase, single-stack, variable-reluctance stepper motor given in Fig. 8.5-1 with θ_{rm} = 0.

Stepper motors are generally designed to operate at current levels which saturate the iron of the machine. Hence, owing to the increased reluctance of the saturated iron, less flux will be circulating around the longer paths through iron than through the shorter paths. Hence, a net mutual flux would exist between stator phases. For the case depicted in Fig. 8.5-4, there would be a net flux in the direction of the positive *bs* axis as a result of the saturation of the iron. Albeit relatively small in amplitude, a mutual inductance does exist in the practical application of single-stack variable-reluctance stepper motors. This complicates the analysis of these devices far beyond that which we care to deal with in this text. Instead, for our first look at stepper motors, we will consider it sufficient to neglect saturation and the mutual coupling it causes in single-stack variable-reluctance stepper motors. A single-stack variable-reluctance stepper motor is shown in Fig. 8.5-5. This device has a 15° step length and is equipped with an integral lead screw for translational motion.

SP8.5-1. Express the number of stator teeth possible for an *N*-phase single-stack variable-reluctance stepper motor. [$ST = n(2N)$, where $n = 1,2,3, ...$]

SP8.5-2. The rotor in Fig. 8.5-1 is replaced by the rotor from Fig. 8.5-3. Determine (a) TP, (b) SL, and (c) the direction of rotation with an *as, bs, cs, as, ...* sequence. [(a) TP = 45°; (b) SL = 15°; (c) cw]

Figure 8.5-5 Single-stack, 15° step, variable-reluctance stepper motor. *Source:* Courtesy of Warner Electric.

8.6 Basic Configuration of Permanent-Magnet Stepper Motors

The permanent-magnet stepper motor is quite common. Actually, it is a permanent-magnet synchronous machine and it may be operated either as a stepping motor or as a continuous-speed device. Here, we will concern ourselves only with its application as a stepping motor since continuous-speed operation is similar to the operation of a permanent magnet ac machine considered in Chapters 4 and 6 [1, 2].

A two-pole two-phase permanent-magnet stepper motor with five rotor teeth is shown in Fig. 8.6-1. Most permanent-magnet stepper motors have more than two poles and more than five rotor teeth; some may have as many as eight poles and as many as 50 rotor teeth. Nevertheless, the elementary device shown in Fig. 8.6-1 is sufficient to illustrate the principle of operation of the permanent-magnet stepper motor. The radial cross-sectional view shown in Fig. 8.6-1b illustrates the permanent magnet which is mounted on the rotor. The permanent magnet magnetizes the iron end caps which are also mounted on the rotor and are slotted to form the rotor teeth. The view looking from left to right at X is shown in Fig. 8.6-1a. Figure 8.6-1c is the view from left to right at Y. The left end cap shown in Fig.8.6-1a is magnetized as a north pole; the right end cap shown in Fig. 8.6-1c is magnetized as a south pole. Note that the rotor teeth of the left end cap are displaced one half a tooth pitch from the teeth on the right end cap. Also, note that the stator windings are wound over the full axial length of the device; a part of the *bs* winding is shown in Fig. 8.6-1b.

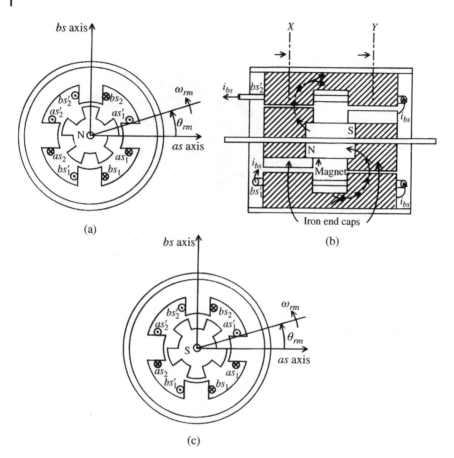

Figure 8.6-1 Two-pole, two-phase, permanent-magnet stepper motor, (a) axial view at *X*; (b) side cross-sectional view; (c) axial view at *Y*.

Let us trace the main path of flux linking the *bs* winding for the rotor position shown in Fig. 8.6-1. This path is depicted by dashed lines in Fig. 8.6-1b; however, it is necessary to visualize the drawing in three dimensions. Flux leaves the left end cap through the rotor tooth at the top of the rotor that is aligned with the stator tooth which has the bs_2 part of the *bs* winding. The flux travels up through the stator tooth in the stator iron. The flux then splits and travels around the circumference of the stator and returns to the south pole of the rotor through the stator tooth, positioned at the bottom in Fig. 8.6-1c, on which the bs_1 winding is wound. The main flux linking the *as* winding for the rotor position shown in Fig. 8.6-1 would enter the stator tooth on which the *as* 1 winding is wound from the rotor tooth on the right of Fig. 8.6-1a. The flux would travel around the circumference of the stator and return to the rotor through the stator pole upon which the as_2 winding is wound, Fig. 8.6-1c.

Stepping action can be explained by first assuming that the bs winding is open-circuited and a constant positive current is flowing in the as winding. As a result of this current, a south pole is established at the stator tooth on which the as_1 winding is wound, and a stator north pole is established at the stator tooth on which the as_2 winding is wound. The rotor would be positioned at $\theta_{rm} = 0$. Now, let us simultaneously deenergize the as winding while energizing the bs winding with a positive current. The rotor will move one step length in the counterclockwise direction. To continue stepping in the counterclockwise direction, the bs winding is deenergized and the as winding is energized with a negative current. That is, counterclockwise stepping occurs with a current sequence of i_{as}, i_{bs}, $-i_{as}$, $-i_{bs}$, i_{as}, ... Clockwise rotation is achieved by i_{as}, $-i_{bs}$, $-i_{as}$, i_{bs},.....

The tooth pitch TP can be calculated from (8.2-1); however, the SL for a permanent-magnet stepper motor cannot be calculated from (8.2-3). As we have mentioned, counterclockwise rotation of the device shown in Fig. 8.6-1 is achieved by a sequence of i_{as}, i_{bs}, $-i_{as}$, $-i_{bs}$, i_{as}, ... We see that it takes four switchings (steps) to advance the rotor one tooth pitch. Thus,

$$TP = 2N\,SL \qquad (8.6\text{-}1)$$

where N is the number of phases. Substituting (8.2-1) into (8.6-1) and solving for SL yields

$$SL = \frac{\pi}{RT\,N} \qquad (8.6\text{-}2)$$

For the device shown in Fig. 8.6-1, $RT = 5$ and $N = 2$. From (8.6-2), $SL = 18°$.

Recall that in the case of variable-reluctance stepper motors, it is unnecessary to reverse the direction of the current in the stator windings to achieve rotation; therefore, the stator voltage source need only be unidirectional. However, in the case of a permanent-magnet stepper motor, it is necessary for the phase currents to flow in both directions to achieve rotation. Generally, stepper motors are supplied from a dc voltage source; hence, the electronic interface between the phase windings and the dc source must be bidirectional; that is, it must have the capability of applying a positive and negative voltage to each phase winding. This requirement markedly increases the cost of the electronic interface and its associated controls relative to a unidirectional source. As an alternative, permanent-magnet stepper motors are often equipped with what is referred to as *bifilar windings*. Rather than only one winding on each stator tooth, there are two identical windings with one wound opposite to the other, each having separate independent external terminals. With this type of winding configuration, the direction of the magnetic field established by the stator windings is reversed, not by changing the direction of the current but by reversing the sense of the winding through which current is flowing. If, for example, the device shown in Fig. 8.6-1 is equipped with bifilar windings, there would be another *as* winding and another *bs* winding with separate, independent, external terminals wound opposite on the stator teeth to the windings shown. Although this increases the size and weight of the stepper motor, it eliminates the need for a bidirectional electronic interface. When this permanent-magnet stepper motor is equipped with bifilar windings as just described, it is (perhaps, inappropriately) called a four-phase device. Actually it has four windings, but it is still a two-phase device magnetically. Although we are not going to consider the bifilar-wound machine in detail, one should be aware of this somewhat ambiguous nomenclature. More specifically, care should be taken when using (8.6-2) to calculate the step length. The number of phases N in (8.6-2) is the number of phases magnetically rather than the number of windings. A cutaway view of a permanent-magnet stepper motor is shown in Fig. 8.6-2.

Figure 8.6-2 Cutaway view of a permanent-magnet stepper motor. *Source:* Courtesy of Sanyo Denki.

SP8.6-1. Consider the device shown in Fig. 8.6-1. The load torque is zero. Initially $i_{as} = I$ and $i_{bs} = 0$. From this condition, the following sequence occurs: $i_{as} = 0$ and $i_{bs} = I$, then $i_{as} = -I$ and $i_{bs} = I$. Determine the initial, intermediate, and final positions. $[\theta_{rm} = 0, 18°, 27°]$

SP8.6-2. A four-pole two-phase permanent-magnet stepper motor has 18 rotor teeth. Calculate TP and SL. $[TP = 20°; SL = 5°]$

8.7 Equations for Permanent-Magnet Stepper Motors

The voltage equations for a two-phase permanent-magnet stepper motor may be written as

$$v_{as} = r_s i_{as} + \frac{d\lambda_{as}}{dt} \tag{8.7-1}$$

$$v_{bs} = r_s i_{bs} + \frac{d\lambda_{bs}}{dt} \tag{8.7-2}$$

In matrix form,

$$\mathbf{v}_{abs} = \mathbf{r}_s \mathbf{i}_{abs} + p\boldsymbol{\lambda}_{abs} \tag{8.7-3}$$

where p is the operator d/dt, and for voltages, currents, and flux linkages

$$(\mathbf{f}_{abs})^T = [f_{as} \quad f_{bs}] \tag{8.7-4}$$

with

$$\mathbf{r}_s = \begin{bmatrix} r_s & 0 \\ 0 & r_s \end{bmatrix} \tag{8.7-5}$$

The flux linkages may be expressed as

$$\lambda_{as} = L_{asas} i_{as} + L_{asbs} i_{bs} + \lambda_{asm} \tag{8.7-6}$$

$$\lambda_{bs} = L_{bsas} i_{as} + L_{bsbs} i_{bs} + \lambda_{bsm} \tag{8.7-7}$$

In matrix form,

$$\boldsymbol{\lambda}_{abs} = \mathbf{L}_s \mathbf{i}_{abs} + \boldsymbol{\lambda}'_m \tag{8.7-8}$$

where

$$\mathbf{L}_s = \begin{bmatrix} L_{asas} & L_{asbs} \\ L_{bsas} & L_{bsbs} \end{bmatrix} \tag{8.7-9}$$

$$\lambda'_m = \begin{bmatrix} \lambda_{asm} \\ \lambda_{bsm} \end{bmatrix} \qquad (8.7\text{-}10)$$

From Fig. 8.6-1, we can write, as a first approximation,

$$\lambda'_m = \lambda'_m \begin{bmatrix} \cos(RT\,\theta_{rm}) \\ \sin(RT\,\theta_{rm}) \end{bmatrix} \qquad (8.7\text{-}11)$$

where λ'_m is the amplitude of the flux linkages established by the permanent magnet as viewed from the stator phase windings. In other words, the magnitude of λ'_m is proportional to the magnitude of the open-circuit sinusoidal voltage induced in each stator phase winding. In (8.7-11),

$$\theta_{rm} = \omega_{rm}t + \theta_{rm}(0) \qquad (8.7\text{-}12)$$

Those who have read Chapter 4 on the brushless dc machines will recognize the similarity in the analysis. The procedure for calculating λ'_m in the case of the stepper motor is identical to that illustrated in Example 4.A.

From the idealized standpoint, the self-inductance of the stator phases of the device shown in Fig. 8.6-1 is constant, and the reluctance seen by the permanent magnet is also constant, independent of rotor position. However, in practice both the self-inductances and the reluctance vary with rotor position due to saturation of the stator iron and the differences from the idealized configuration which occur when shaping the poles. We shall disregard these departures from the idealized case and assume constant self-inductances and a constant reluctance seen by the permanent magnet independent of rotor position. When doing so, we are neglecting the reluctance torques caused by variation in self-inductances and the permanent magnet, both of which attempt to place the rotor in its minimum-reluctance position. The latter torque is often referred to as the *detent* or *retention torque,* since it exists whether or not the stator windings are excited, and, if the load torque is not too large, this detent torque will preserve the rotor position during a power failure. Nevertheless, the reluctance torques are small relative to the torque produced by the interaction of the permanent magnet and the stator currents and, although we are not looking at the complete picture when we neglect the reluctance torques, this approximation is certainly adequate for our first look at the permanent-magnet stepper motor.

With the assumption of constant self-inductances, we can write

$$L_{asas} = L_{ls} + L_{ms} = L_{ss} \qquad (8.7\text{-}13)$$

$$L_{bsbs} = L_{ls} + L_{ms} = L_{ss} \qquad (8.7\text{-}14)$$

Following a line of reasoning similar to that used in the case of the single-stack variable-reluctance stepper motor, it can be shown that stator mutual inductances do not exist if saturation is neglected. Thus,

$$\mathbf{L}_s = \begin{bmatrix} L_{ss} & 0 \\ 0 & L_{ss} \end{bmatrix} \tag{8.7-15}$$

An expression for the electromagnetic torque may be obtained by taking the partial derivative of the coenergy with respect to θ_{rm}. Since the stator mutual inductances are zero, the coenergy may be expressed as

$$W_c = \frac{1}{2} L_{asas} i_{as}^2 + \frac{1}{2} L_{bsbs} i_{bs}^2 + \lambda_{asm} i_{as} + \lambda_{bsm} i_{bs} + W_{pm} \tag{8.7-16}$$

where L_{asas} and L_{bsbs} are given by (8.7-13) and (8.7-14), respectively, and λ_{asm} and λ_{bsm} are given by (8.7-11). The term W_{pm} is related to the energy associated with the permanent magnet. Since we are neglecting variations in the self-inductances and in W_{pm} taking the partial derivative of W_c with respect to θ_{rm} yields

$$T_e = -RT\ \lambda'_m [i_{as} \sin{(RT\ \theta_{rm})} - i_{bs} \cos{(RT\ \theta_{rm})}] \tag{8.7-17}$$

The terms of (8.7-17) are plotted in Fig. 8.7-1, wherein it is assumed that constant currents are present in both phase windings. Each term of (8.7-17) is identified in Fig. 8.7-1. In particular, $\pm T_{eam}$ is the torque due to the interaction of the

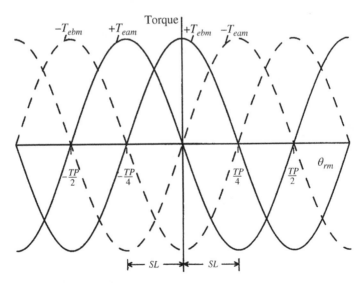

Figure 8.7-1 Plot of T_e versus θ_{rm} for a permanent-magnet stepper motor with constant phase currents.

permanent magnet and $\pm i_{as}$, and $\pm T_{ebm}$ are due to the interaction of the permanent magnet and $\pm i_{bs}$.

The reluctance of the permanent magnet is large, approaching that of air. Since the flux established by the phase currents flows through the magnet, the reluctance of the flux path is relatively large. Hence, the variation in the reluctance due to rotation of the rotor is small and, consequently, the amplitudes of the reluctance torques are small relative to the torque produced by the interaction between the permanent magnet and the phase currents. For this reason, the reluctance torques are generally neglected, as we have done here, and the self-inductances are assumed to be constant. Therefore, the voltage equations for the permanent-magnet stepper motor become those of the permanent-magnet ac machine which we considered in Chapters 4 and 6, except for the use of θ_{rm} instead of θ_r and the difference in the referencing of θ_{rm} (to the minimum reluctance rather than the maximum reluctance).

Although a discussion of the stepping action of a permanent-magnet stepper motor using the steady-state torque-angle characteristics is appropriate, this explanation would be essentially a repeat of that given in Section 8.4 for the variable-reluctance devices. We will not do this; instead, we will ask a few questions to help emphasize this similarity. Also, we can use reference frame theory to express the equations in the rotor reference frame; however, this is similar to that covered in Chapters 4 and 6 and it is also carried out in [2].

SP8.7-1. Express λ'_{asm} in terms of SL rather than RT. Determine the number of step lengths in a period for the device shown in Fig. 8.6-1. $\{\lambda'_{asm} = \lambda'_m \cos[\pi/(SL\,N)]\theta_{rm}; 4\}$

SP8.7-2. Consider Fig. 8.7-1. The load torque is zero. Initially, $i_{as} = I$, then $i_{bs} = 0$ and $i_{bs} = -3I$, and, finally, $i_{as} = -\dfrac{1}{2}I$ and $i_{bs} = 0$. Determine the three positions. $[\theta_{rm} = 0, -TP/4, -TP/2]$

8.8 Problems

1 Sketch the configuration of a two-pole four-stack variable-reluctance stepper motor with two rotor teeth. Use *as, bs, cs,* and *ds* to denote the phase windings. Calculate *TP, SL,* and give the excitation sequence for ccw rotation.

2 For Prob. 1, express the self-inductances and the torque using *SL* in the arguments.

3 The four-pole three-stack variable-reluctance stepper motor shown in Fig. 8.2-3 is to be operated at a continuous speed of 30 rad/s. Neglect electrical transients and sketch the current i_{as} indicating the time it is zero and nonzero.

4 A four-pole five-stack variable-reluctance stepper motor has eight rotor teeth as the rotor shown in Fig. 8.2-4. Its magnetic axes are arranged *as, bs, cs, ds,* and *es,* in the counterclockwise direction. Express the self-inductances with the constant angular displacement in terms of step length.

5 Express the self-inductances for the single-stack variable-reluctance stepper motor shown in Fig. 8.5-1 with the constant angular displacement in terms of step length.

6 A two-phase permanent-magnet stepper motor has 50 rotor teeth. When the rotor is driven by an external mechanical source at ω_{rm}=100 rad/s, the measured open-circuit phase voltage is 25 V, peak to peak. Calculate λ_m and *SL*. If i_{as}=1 A, $i_{bs}=0$, express T_e.

7 Consider the two-phase permanent-magnet stepper motor of Fig. 8.6-1. Sketch i_{as} and i_{bs} versus time for the excitation sequence $i_{as}, i_{bs}, -i_{as}, -i_{bs}, i_{as}, ...$ Denote the time between steps as T_s and the stepping rate as $f_s = 1/T_s$. Establish a relationship between the fundamental frequency (ω_e) of i_{as} and i_{bs}, and the stepping rate f_s. Relate ω_{rm} to ω_e and to f_s.

8 A two-phase permanent-magnet stepper motor has 50 rotor teeth. The parameters are $\lambda'_m = 0.00226$ V · s/rad, $r_s = 10\,\Omega$, and L_{ss}=1.1 mH. The applied stator voltages form a balanced two-phase set with V_s=10 V, ω_e=314 rad/s. Establish the steady-state rotor speed ω_{rm} and the maximum electromagnetic torque T_{eM} that can be developed at this speed.

References

1 P. P. Acarnley, *Stepping Motors: A Guide to Modern Theory and Practice*, Peter Pereginus Ltd. for the Institution of Electrical Engineers; Southgate House, Stevenage, Herts, SOl IHQ, England, 1984.

2 P. C. Krause, O. Wasynczuk, S. D. Pekarek, and T. O'Connell, *Electromechanical Motion Devices*, 3rd Ed. Wiley, and IEEE Press, Hoboken, NJ, USA, 2020.

Appendix A

Abbreviations, Constants, Conversions, and Identities

Term	Abbreviation
alternating	ac
ampere	A
ampere-turn	At
coulomb	C
direct current	dc
electromotive force	emf
foot	ft
gauss	G
gram	g
henry	H
hertz	Hz
horsepower	hp
inch	in
joule	J
kilogram	kg
kilovar	kvar
kilovolt	kV
kilovoltampere	kVA
kilowatt	kW
magnetomotive force	mmf

Introduction to Modern Analysis of Electric Machines and Drives, First Edition.
Paul C. Krause and Thomas C. Krause.
© 2023 The Institute of Electrical and Electronics Engineers, Inc.
Published 2023 by John Wiley & Sons, Inc.

Term	Abbreviation
maxwell	Mx
megawatt	MW
meter	m
microfarad	μF
millihenry	mH
newton	N
newton meter	N · m
oersted	Oe
pound	lb
poundal	pdl
power factor	pf
pulse-width modulation	PWM
radian	rad
revolution per minute	r/min (rpm)
root mean square	rms
second	s
voltampere reactive	var
volt	V
voltampere	VA
watt	W
weber	Wb

Constants and Conversion Factors

permeability of free space	$\mu_0 = 4\pi \times 10^{-7}\,\text{Wb/A} \cdot \text{m}$
permittivity of free space	$\varepsilon_0 = 8.854 \times 10^{-12}\text{C}^2/\text{N} \cdot \text{m}^2$
acceleration of gravity	$g = 9.807\,\text{m/s}^2$
length	$1\,\text{m} = 3.218\,\text{ft} = 39.37\,\text{in}$
mass	$1\,\text{kg} = 0.0685\,\text{slug} = 2.205\,\text{lb (mass)}$
force	$1\,\text{N} = 0.225\,\text{lb} = 3.6\,\text{oz}$
torque	$1\,\text{N} \cdot \text{m} = 0.738\,\text{lb} \cdot \text{ft}$
energy	$1\,\text{J (W} \cdot \text{s)} = 0.738\,\text{lb} \cdot \text{ft}$
power	$1\,\text{W} = 1.341 \times 10^{-3}\,\text{hp}$
moment of inertia	$1\,\text{kg} \cdot \text{m}^2 = 0.738\,\text{slug} \cdot \text{ft}^2 = 23.7\,\text{lb} \cdot \text{ft}^2$
magnetic flux density	$1\,\text{Wb/m}^2 = 10,000\,\text{G} = 64.5\,\text{klines/in}^2$
magnetizing force	$1\,\text{At/m} = 0.0254\,\text{At/in} = 0.0126\,\text{Oe}$

Trigonometric Identities

(I-1) $e^{j\alpha} = \cos\alpha + j\sin\alpha$

(I-2) $a\cos x + b\sin x = \sqrt{a^2 + b^2}\,\cos(x + \phi)$ $\phi = \tan^{-1}(-b/a)$

(I-3) $\cos^2 x + \sin^2 x = 1$

(I-4) $\sin 2x = 2\sin x \cos x$

(I-5) $\cos 2x = \cos^2 x - \sin^2 x = 2\cos^2 x - 1 = 1 - 2\sin^2 x$

(I-6) $\cos x\,\cos y = \dfrac{1}{2}\cos(x+y) + \dfrac{1}{2}\cos(x-y)$

(I-7) $\sin x\,\sin y = \dfrac{1}{2}\cos(x-y) - \dfrac{1}{2}\cos(x+y)$

(I-8) $\sin x\,\cos y = \dfrac{1}{2}\sin(x+y) + \dfrac{1}{2}\sin(x-y)$

(I-9) $\cos(x \pm y) = \cos x\,\cos y \mp \sin x\,\sin y$

(I-10) $\sin(x \pm y) = \sin x\,\cos y \pm \cos x\,\sin y$

(I-11) $\cos^2 x + \cos^2\left(x - \dfrac{2}{3}\pi\right) + \cos^2\left(x + \dfrac{2}{3}\pi\right) = \dfrac{3}{2}$

(I-12) $\sin^2 x + \sin^2\left(x - \dfrac{2}{3}\pi\right) + \sin^2\left(x + \dfrac{2}{3}\pi\right) = \dfrac{3}{2}$

(I-13) $\sin x\,\cos x + \sin\left(x - \dfrac{2}{3}\pi\right)\cos\left(x - \dfrac{2}{3}\pi\right) + \sin\left(x + \dfrac{2}{3}\pi\right)\cos\left(s + \dfrac{2}{3}\right) = 0$

(I-14) $\cos x + \cos\left(x - \dfrac{2}{3}\pi\right) + \cos\left(x + \dfrac{2}{3}\pi\right) = 0$

(I-15) $\sin x + \sin\left(x - \dfrac{2}{3}\pi\right) + \sin\left(x + \dfrac{2}{3}\pi\right) = 0$

(I-16) $\sin x\,\cos y + \sin\left(x - \dfrac{2}{3}\pi\right)\cos\left(y - \dfrac{2}{3}\pi\right) + \sin\left(x + \dfrac{2}{3}\pi\right)\cos\left(y + \dfrac{2}{3}\pi\right) = \dfrac{3}{2}\cos(x - y)$

(I-17) $\sin x\,\sin y + \sin\left(x - \dfrac{2}{3}\pi\right)\sin\left(y - \dfrac{2}{3}\pi\right) + \sin\left(x + \dfrac{2}{3}\pi\right)\sin\left(y + \dfrac{2}{3}\pi\right) = \dfrac{3}{2}\cos(x - y)$

(I-18) $\cos x\,\sin y + \cos\left(x - \dfrac{2}{3}\pi\right)\sin\left(y - \dfrac{2}{3}\pi\right) + \cos\left(x + \dfrac{2}{3}\pi\right)\sin\left(y + \dfrac{2}{3}\pi\right) = -\dfrac{3}{2}\sin(x - y)$

(I-19) $\cos x\,\cos y + \cos\left(x - \dfrac{2}{3}\pi\right)\sin\left(y - \dfrac{2}{3}\pi\right) + \cos\left(x + \dfrac{2}{3}\pi\right)\sin\left(y + \dfrac{2}{3}\pi\right) = \dfrac{3}{2}\cos(x - y)$

(I-20) $\sin x\,\cos y + \sin\left(x + \dfrac{2}{3}\pi\right)\cos\left(y - \dfrac{2}{3}\pi\right) + \sin\left(x + \dfrac{2}{3}\pi\right)\cos\left(y + \dfrac{2}{3}\pi\right) = \dfrac{3}{2}\sin(x - y)$

(I-21) $\sin x\,\sin y + \sin\left(x + \dfrac{2}{3}\pi\right)\cos\left(y - \dfrac{2}{3}\pi\right) + \sin\left(x - \dfrac{2}{3}\pi\right)\sin\left(y + \dfrac{2}{3}\pi\right) = -\dfrac{3}{2}\cos(x + y)$

(I-22) $\cos x\,\sin y + \cos\left(x + \dfrac{2}{3}\pi\right)\sin\left(y - \dfrac{2}{3}\pi\right) + \cos\left(x - \dfrac{2}{3}\pi\right)\sin\left(y + \dfrac{2}{3}\pi\right) = \dfrac{3}{2}\sin(x + y)$

(I-23) $\cos x\,\cos y + \cos\left(x + \dfrac{2}{3}\pi\right)\cos\left(y - \dfrac{2}{3}\pi\right) + \cos\left(x - \dfrac{2}{3}\pi\right)\cos\left(y + \dfrac{2}{3}\pi\right) = \dfrac{3}{2}\cos(x + y)$

Epilogue

The goal of this book is to introduce the undergraduate electrical engineer, who has an interest in electric power and/or drives areas, to modern analysis of electric machines. With the emergence of electric drives, an interested person with a bachelor's degree in electrical engineering should have a working knowledge of reference frame theory. This is no longer just a graduate subject. This book sets forth reference frame theory and the arbitrary reference frame as the backbone of the analysis of alternating current machines. Unlike other undergraduate books where reference frame theory is either deemphasized or optional, here, reference frame theory is the basis of analysis. The student must follow derivations using reference frame theory which leads to a clearer understanding of the variables that can be controlled for different applications and sets the stage for the design of controls for new applications.

In the first chapter, some of the classic tools for machine analysis were set forth. The stator is the same for most induction and synchronous machines. Thus, the stator was analyzed once in the second chapter and this derivation is not repeated. It was shown that the transformation used in reference frame theory comes from the equation for Tesla's rotating magnetic field. This transformation provides the circuits in the reference frame of interest with the appropriate voltages, currents, and flux linkages that produce/portray Tesla's rotating magnetic field as viewed from that reference frame. This fact makes reference frame theory and the transformation intuitive and understandable. Analysis and discussions explained that for reference frame speeds less than synchronous speed, the rotating magnetic field rotates counterclockwise for an *abc* sequence. For reference frame speeds above synchronous speed, the rotating magnetic field rotates clockwise.

The induction machine, which is the workhouse of the industry, was analyzed next. It was shown that the squirrel-cage rotor, which is common in most singly fed induction machines, can be analyzed as a wound rotor similar in structure to the

Introduction to Modern Analysis of Electric Machines and Drives, First Edition.
Paul C. Krause and Thomas C. Krause.
© 2023 The Institute of Electrical and Electronics Engineers, Inc.
Published 2023 by John Wiley & Sons, Inc.

stator. The rotor analysis parallels previous work with the stator. The arbitrary reference frame was shown to be particularly useful for induction machines and yielded machine models for dynamic simulations and the steady-state single-phase equivalent circuit. Simulations viewed from the stationary, rotor, and synchronous reference frames were shown and discussed.

Two types of synchronous machines were considered: the permanent-magnet synchronous machine commonly used in brushless dc drives and the wound-rotor synchronous generator. The first type is becoming the motor of choice for many applications and the second is the main source of electric grid power. In a brushless dc drive, the stator of the permanent magnet synchronous machine is driven by a variable-frequency inverter. The frequency of the inverter is controlled to be the electrical angular frequency of the rotor. This produces a torque speed characteristic like a brushed permanent magnet dc machine. The synchronous generator has a rotor with a single-phase field winding and damper windings. Using work that we had done earlier in the text, we wrote out the equations for the synchronous generator from the equivalent circuit. Operating characteristics of the synchronous generator were shown, and discussions included an introduction to transient stability which is important for the power systems engineer.

The next chapter included a brief treatment of the brushed permanent magnet dc machine, primarily for comparison with the ac motor drives which were covered in the following chapter. Therein, the brushless dc drive and field-oriented induction motor were considered. In the case of the brushless dc drive, the equations were established for normal operation as well as maximum torque per volt and per ampere. Simulations and phasor diagrams demonstrated the differences of the operating modes. In the case of the induction motor drive, the field-oriented equations were established, and torque speed characteristics shown.

The final two chapters covered single-phase induction motors and stepper motors, respectively. These chapters are informative for students interested in the drives area. The theory of symmetrical components helped analyze the single-phase induction motor. In the stepper motors chapter, the variable-reluctance stepper motor and permanent magnet stepper motor were described and analyzed.

We hope this book helps students understand and appreciate electric machines and drives. We feel this is the way machines should and will be taught in the future. Hopefully this is a step in the right direction.

The following is a list of books published by Wiley and IEEE Press that provide advanced reading.

1. *Analysis of Electric Machinery*
   Paul C. Krause, Oleg Wasynczuk, and S. D. Sudhoff

2. *Principles of Electric Machines with Power Electronic Applications,*
   Second Edition
   Mohamed E. El-Hawary
3. *Pulse Width Modulation for Power Converters: Principles and Practice*
   D. Grahame Holmes and Thomas A. Lipo
4. *Analysis of Electric Machinery and Drive Systems,* Second Edition
   Paul C. Krause, Oleg Wasynczuk, and Scott D. Sudhoff
5. *Control of Electric Machine Drive Systems*
   Seung-Ki Sul
6. *Power Conversion and Control of Wind Energy Systems*
   Bin Wu, Yongqiang Lang, Navid Zargari, and Samir Kouro
7. *Doubly Fed Induction Machine: Modeling and Control for Wind Energy Generation*
   Gonzalo Abad, Jesús López, Miguel Rodrigues, Luis Marroyo, and Grzegorz Iwanski
8. *Electromechanical Motion Devices,* Second Edition
   Paul Krause, Oleg Wasynczuk, and Steven Pekarek
9. *Analysis of Electric Machinery and Drive Systems,* Third Edition
   Paul Krause, Oleg Wasynczuk, Scott Sudhoff, and Steven Pekarek
10. *Power Magnetic Devices: A Multi-Objective Design Approach*
    S. D. Sudhoff
11. *Modeling and High-Performance Control of Electric Machines*
    John Chiasson
12. *Model Predictive Control of Wind Energy Conversion Systems*
    Venkata Yaramasu and Bin Wu
13. *High-Power Converters and AC Drives,* Second Edition
    Bin Wu and Mehdi Narimani
14. *Current Signature Analysis for Condition Monitoring of Cage Induction Motors: Industrial Application and Case Histories*
    William T. Thomson and Ian Culburt
15. *Introduction to AC Machine Design*
    Thomas A. Lipo
16. *Multiphysics Simulation by Design for Electrical Machines, Power Electronics, and Drives*
    Marius Rosu, Ping Zhou, Dingshen Lin, Dan Ionel, Mircea Popescu, Frede Blaabjerg, Vandana Rallabandi, and David Staton
17. *Advanced Control of Doubly Fed Induction Generator for Wind Power Systems*
    Dehong Xu, Frede Blaabjerg, Wenjie Chen, and Nan Zhu
18. *Electromechanical Motion Devices: Rotating Magnetic Field-Based Analysis and Online Animations,* Third Edition
    Paul Krause, Oleg Wasynczuk, Steven D. Pekarek, and Timothy O'Connell

19. *Reference Frame Theory: Development and Applications*
Paul Krause

Here are a few papers for advanced reading:

1. T. M. Rowan and R. J. Kerkman, "A New Synchronous Current Regulator and an Analysis of Current-Regulated PWM Inverters," *IEEE Transactions on Industry Applications*, Vol. IA-22, no. 4, July 1986, pp. 678–690. doi: 10.1109/TIA.1986.4504778.
2. S. R. Shaw and S. B. Leeb, "Identification of Induction Motor Parameters From Transient Stator Current Measurements," *IEEE Transactions on Industrial Electronics*, Vol. 46, no. 1, February 1999, pp. 139–149. doi: 10.1109/41.744405.
3. F. Therrien, L. Wang, J. Jatskevich, and O. Wasynczuk, "Efficient Explicit Representation of AC Machines Main Flux Saturation in State-Variable-Based Transient Simulation Packages," *IEEE Transactions on Energy Conversion*, Vol. 28, no. 2, June 2013, pp. 380–393. doi: 10.1109/TEC.2013.2245332.
4. R. Li and D. Xu, "A Zero-Voltage Switching Three-Phase Inverter," *IEEE Transactions on Power Electronics*, Vol. 29, no. 3, March 2014, pp. 1200–1210. doi: 10.1109/TPEL.2013.2260871.
5. Z. Zhang, F. Wang, L. M. Tolbert, B. J. Blalock, and D. J. Costinett, "Evaluation of Switching Performance of SiC Devices in PWM Inverter-Fed Induction Motor Drives," *IEEE Transactions on Power Electronics*, Vol. 30, no. 10, October 2015, pp. 5701–5711. doi: 10.1109/TPEL.2014.2375827.
6. P. C. Krause, O. Wasynczuk, T. C. O'Connell, and M. Hasan, "Tesla's Contribution to Electric Machine Analysis," *IEEE Transactions on Energy Conversion*, Vol. 32, no. 2, June 2017, pp. 591–598. doi: 10.1109/TEC.2016.2640018.
7. F. Kutt, M. Michna, and G. Kostro, "Multiple Reference Frame Theory in the Synchronous Generator Model Considering Harmonic Distortions Caused by Nonuniform Pole Shoe Saturation," *IEEE Transactions on Energy Conversion*, Vol. 35, no. 1, March 2020, pp. 166–173. doi: 10.1109/TEC.2019.2951858.
8. Y. Xiao, B. Fahimi, M. A. Rotea, and Y. Li, "Multiple Reference Frame-Based Torque Ripple Reduction in DFIG-DC System," *IEEE Transactions on Power Electronics*, Vol. 35, no. 5, May 2020, pp. 4971–4983. doi: 10.1109/TPEL.2019.2941957.
9. D. B. Rathnayake et al., "Grid Forming Inverter Modeling, Control, and Applications," *IEEE Access*, Vol. 9, 2021, pp. 114781–114807. doi: 10.1109/ACCESS.2021.3104617.
10. C. J. O'Rourke, M. M. Qasim, M. R. Overlin, and J. L. Kirtley, "A Geometric Interpretation of Reference Frames and Transformations: dq0, Clarke, and Park," *IEEE Transactions on Energy Conversion*, Vol. 34, no. 4, December 2019, pp. 2070–2083. doi: 10.1109/TEC.2019.2941175.

Index

Introduction to Modern Analysis of Electric Machines and Drives, First Edition.
Paul C. Krause and Thomas C. Krause.
© 2023 The Institute of Electrical and Electronics Engineers, Inc.
Published 2023 by John Wiley & Sons, Inc.

IEEE Press Series on Power and Energy Systems

Series Editor: Ganesh Kumar Venayagamoorthy, Clemson University, Clemson, South Carolina, USA.

The mission of the IEEE Press Series on Power and Energy Systems is to publish leading-edge books that cover a broad spectrum of current and forward-looking technologies in the fast-moving area of power and energy systems including smart grid, renewable energy systems, electric vehicles and related areas. Our target audience includes power and energy systems professionals from academia, industry and government who are interested in enhancing their knowledge and perspectives in their areas of interest.

Printed and bound by CPI Group (UK) Ltd, Croydon, CR0 4YY

16/04/2025

14658416-0002